Optimal High-Throughput Screening

This concise, self-contained, and cohesive book focuses on commonly used and recently developed methods for designing and analyzing high-throughput screening (HTS) experiments from a statistically sound basis. Combining ideas from biology, computing, and statistics, the author explains experimental designs and analytic methods that are amenable to rigorous analysis and interpretation of RNAi HTS experiments. The opening chapters are carefully presented to be accessible both to biologists with training only in basic statistics and to computational scientists and statisticians with basic biological knowledge. Biologists will see how new experimental designs and rudimentary data-handling strategies for RNAi HTS experiments can improve their results, whereas analysts will learn how to apply recently developed statistical methods to interpret HTS experiments.

Dr. Xiaohua Douglas Zhang is an associate director at Merck Research Laboratories. He has worked on data analysis for genome-wide RNAi research and microarrays in drug discovery and development for various diseases for many years. He has continuously developed novel analytic methods and experimental designs for quality control and hit selection in genome-scale RNAi research. Dr. Zhang and his colleagues have published many articles in various peer-reviewed journals, including *Cell Host & Microbe, Nucleic Acids Research, Bioinformatics, Genetic Epidemiology, Journal of Biological Chemistry, Pharmacogenomics, Genomics,* and *Journal of Biomolecular Screening,* among many others.

Optimal High-Throughput Screening

Practical Experimental Design and Data Analysis
for Genome-Scale RNAi Research

Xiaohua Douglas Zhang

Merck Research Laboratories

CAMBRIDGE
UNIVERSITY PRESS

CAMBRIDGE UNIVERSITY PRESS
Cambridge, New York, Melbourne, Madrid, Cape Town,
Singapore, São Paulo, Delhi, Tokyo, Mexico City

Cambridge University Press
32 Avenue of the Americas, New York, NY 10013-2473, USA

www.cambridge.org
Information on this title: www.cambridge.org/9780521734448

First published 2011

Printed in the United States of America

A catalog record for this publication is available from the British Library.

Library of Congress Cataloging in Publication data

Zhang, Xiaohua Douglas, 1966– author.
Optimal high-throughput screening : practical experimental design and data analysis for
genome-scale RNAi research / Xiaohua Douglas Zhang, Merck Research Laboratories.
    p. ; cm.
Includes bibliographical references and index.
ISBN 978-0-521-51771-3 (hardback) – ISBN 978-0-521-73444-8 (paperback)
1. High throughput screening (Drug development)   2. Small interfering RNA.
3. Experimental design.   I. Title.
[DNLM: 1. High-Throughput Screening Assays – methods.   2. RNA Interference.   3. Research Design.
4. Statistics as Topic – methods.   QV 778]
RS419.5.Z43   2011
615′.19–dc22   2010051114

ISBN 978-0-521-51771-3 Hardback
ISBN 978-0-521-73444-8 Paperback

100661656 1

Contents

Color plates follow page 110.

Preface

In 2000, scientists triumphantly announced they had deciphered the human genome, the blueprint for human life; in 2001, almost the entire human genome sequence became principally known. In 2003, the Human Genome Project was completed. By laying out in order the 3.2 billion units of our DNA, researchers sparked a firestorm of discovery and an explosion of genomic knowledge, which have been accompanied by rapidly emerging novel genomic technologies, including micro-arrays, whole-genome *single nucleotide polymorphism* (SNP) chips, *RNA interference* (RNAi) *high-throughput screening* (HTS), and so forth. All these have launched a new era – the genomic revolution era, which offers us boundless potential and great promise. Foremost are prospects in health, ranging from discovering cures for cancer to developing personalized medical products for individuals. The success in applying the "genomic revolution" to the discovery and development of new medical products largely depends on our ability to understand gene and gene interactions associated with drug response and disease. RNAi is a natural mechanism for gene silencing that can be harnessed to reveal information about gene function [48], leading to advances not only in drug target identification and validation, but also in the development of a potentially whole new class of therapeutic agents based on RNAi [24].

RNAi was first characterized as post-transcriptional gene silencing in petunia [109]. Later studies in *Caenorhabditis elegans* revealed that the interference with gene function was triggered by the presence of double-stranded RNA [48]. Exogenous delivery of double-stranded RNA was then developed as an experimental tool for functional genomics: first, in *C. elegans* and *Drosophila* and later, in mammalian cell culture systems, when it was discovered that delivery of short double-stranded RNA oligonucleotides triggers RNAi without inducing the interferon response [51]. This type of RNA oligonucleotide is thus called *small interfering RNA* (siRNA). The development of algorithms for siRNA design that produce a potent and selective knockdown of targeted genes has led to a great deal of interest in using siRNAs to elucidate gene function and identify novel targets for drug discovery. The importance of RNAi was further recognized when the Nobel Prize in medicine and physiology was awarded to A. Fire and C. C. Mello in 2006 for their research in this field [48]. The application of genome-scale RNAi relies on the development of RNAi HTS technology.

RNAi HTS is broadly used in the identification of genes associated with specific biological phenotypes. This technology has been hailed as the second genomics wave, following the first genomics wave of gene expression microarray and single nucleotide polymorphism discovery platforms [101]. Before the emergence of RNAi HTS, compound HTS (which allows rapid screening of large collections of compounds consisting of small molecules) had been widely used in the pharmaceutical industry. As in any high-throughput platform, one of the most fundamental challenges in RNAi/compound HTS is gleaning biological significance from large volumes of data that rely on the development and adoption of appropriate statistical designs and analytic methods for quality control and hit selection [43].

Merck has applied extensive effort into RNAi research. It purchased Sirna Therapeutics for $1.1 billion in October 2006 [186] and has one of the largest labs for conducting genomewide RNAi research and compound HTS. Since early 2005, I have led data analysis in RNAi HTS projects at Merck and have continuously developed and adopted experimental designs and analytic methods for genome-scale RNAi research, including novel analytic methods for quality control and hit selection, which has allowed my colleagues and me to publish multiple articles on genome-scale RNAi research [28;45;86;161–180;182;183]. In 2005, I gave presentations on statistical methods for RNAi HTS at the Joint Statistical Meetings (Minneapolis, Minnesota) and the RNAi Meeting (Cold Spring Harbor, New York). Since then, I have given invited presentations and seminars at, among others, the 2006 International Conference on Bioinformatics and Computational Biology (Las Vegas, Nevada); 2006 Joint Statistical Meetings (Seattle, Washington); 2007 Seminars of Institute of Microbiology, Chinese Academy of Sciences (Beijing, China); 2007 International Conference of Bioinformatics (Hong Kong, China); 2007 Joint Statistical Meetings (Salt Lake City, Utah); 2008 Department of Statistics Seminar, Temple University (Philadelphia, Pennsylvania); 2008 RNAi and miRNA World Congress (Boston, Massachusetts); and 2009 World Pharmaceutical Congress (Philadelphia, Pennsylvania).

During my presentations, the following questions are usually asked by members of the audience: "As a statistician, I want to know about recently developed statistical methods and analytic tools in genome-scale RNAi research. Could you suggest a book in this field?" "As a biologist, although I have some knowledge of statistics, I do not have systematic training in this field. I am very interested in reading a book that introduces analytic tools in genome-scale RNAi research. Do you know of any book that describes the necessary and basic statistics knowledge in genome-scale RNAi research so that I can apply it to our experiments without knowing much details about statistics?" Or, "I am a graduate student. I am very interested in data analysis in genome-scale RNAi research and am eager to work in this area in the near future whenever possible. Could you tell me which book will describe the necessary scientific knowledge and prepare me well for data analysis in this area?" Obviously, there is a demand for a self-contained and cohesive book about data analysis on genome-scale RNAi research. However, to my knowledge, such a book had yet to be written. The demands from the audiences and the need to promote genome-scale RNAi research propelled me to write such a book, in which I describe and present scientific knowledge and recently developed analytic methods and applications based on my experience in developing them and analyzing many genome-scale RNAi projects in the pharmaceutical industry.

Audience

In genome-scale RNAi research, it takes an ongoing dialog and a two-way flow of information and ideas between biologists and computational scientists, including statisticians, to develop experimental designs and analytic methods that are amenable to rigorous analysis and interpretation of RNAi HTS experiments. It has been recognized that biologists have an unfortunate tendency to "plug and play" with analytic methods without understanding the underlying principles, resulting in the misuse of otherwise effective strategies. Thus, at this time, most biologists depend on their computational colleagues for the development of data analysis methods, and most computational scientists depend on their biology colleagues to perform experiments that address important biological questions and to generate data [5]. Meanwhile, some people believe that, soon, if a scientist does not understand some statistics or rudimentary data-handling technologies, he or she may not be considered a true molecular biologist and thus will simply become a dinosaur [43]. On the other hand, to perform appropriate and effective data analysis, computational scientists need to know the details about recently developed methods and understand basic biological processes and technologies in HTS experiments.

Considering the needs of both biologists and computational scientists, this book has two major goals: 1) to help biologists who have limited training in statistics understand experiment designs, recently developed statistical methods, and rudimentary data-handling strategies for RNAi HTS experiments; and 2) to help computational scientists grasp recently developed statistical methods and common analytic tools and then be able to use them for analyzing data in HTS experiments. It is also suitable for graduate students (and perhaps undergraduate students) of biology or computational science who want to learn data analysis in HTS technologies. The first part of this book should be generally comprehensible to a biologist with training in basic statistics, as well as to a computational scientist with basic biological knowledge; the second part of this book should be comprehensible to a computational scientist with a master's degree or equivalent in statistics/biostatistics. The analytic methods presented in this book should also be suitable for biologists/chemists and computational scientists working in any other HTS, including small-molecule HTS experiments.

Content and General Outline

This is a concise, self-contained, and cohesive book focusing on commonly used and recently developed methods for designing an HTS experiment from a statistically sound basis and for analyzing data from the experiment. The topics of this book reflect my personal experiences and biases in designing and analyzing HTS data. A significant portion of the book is built on material from articles that my colleagues and I have written, talks I have presented at multiple conferences, and my unpublished observations. Although I have tried to quote relevant literature, I may have missed some related references.

Chapter 1 presents an introduction to RNAi and HTS technologies and a description of a typical RNAi HTS project in the pharmaceutical industry. Chapter 2 provides experimental designs for genome-scale RNAi screens, which include siRNA designs, control designs, plate designs, designs for siRNA delivery and optimization of transfection, and sample size designs. Chapter 3 discusses how to display data in order to identify potential systematic errors, how to determine data transformation, and how to adjust for identified systematic errors. In Chapter 4, I present both biological processes and analytic methods for quality control and demonstrate how to apply them in HTS experiments. In an RNAi HTS, a primary goal is to select siRNAs with a desired size of inhibition or activation effect. An siRNA with a desired size of effect in an HTS screen is called a hit. The process of selecting hits is called hit selection. The analytic methods for hit selection in the screens without replicates differ from those with replicates. Therefore, I present them separately: without replicates in Chapter 5 and with replicates in Chapter 6. In Chapters 5 and 6, I explore classic analytic methods, including z-score method and t-test; describe recently developed methods, including robust methods, error control methods, and methods based on *strictly standardized mean difference* (SSMD) for hit selection; briefly introduce analytic methods for addressing off-target effects; and illustrate how to use them in RNAi HTS experiments. Sample size consideration for hit selection is also explored in Chapters 5 and 6. Chapters 1 through 6, which are presented in Part I of this book, are written so that a scientist with training in basic statistics can understand them. In each of these six chapters, I provide strategies on when and how to apply experimental designs and analytic methods in practical RNAi HTS experiments.

In contrast, Chapters 7 and 8, which are presented in Part II, describe and derive recently developed analytic methods from a solid statistical foundation and thus require the reader to have systematic training in statistics. Specifically, in Chapter 7, I present newly developed statistical methods for comparing groups, including contrast variable, c^+-probability, d^+-probability, *standardized mean of a contrast variable* (SMCV), and their statistical estimation and inference; derive SMCV-based criteria for assessing the strength of group comparisons; and extend the concepts to the settings of multifactor *analysis of variance* (ANOVA). Chapter 7 builds a strong theoretical base for newly developed statistical methods for assessing the size of siRNA effects and for addressing off-target effects. In Chapter 8, I describe and derive newly developed statistical methods for assessing the size of siRNA effects, which includes SSMD and associated error-control methods, such as false discovery rate, false non-discovery rate, p-value, p^*-value, q-value, and q^*-value for hit selection in RNAi HTS experiments. In Chapter 8, I also elaborate on analytic methods adjusting for off-target effects. R functions for most analytic methods in this book will be formed into an R library and should be submitted to Bioconductor (www.bioconductor.org).

Acknowledgments

First, I would like to thank Senior Editor Lauren Cowles of Cambridge University Press for her continuous work in motivating and encouraging me to write this book; otherwise, I might not have considered this challenge. I am deeply indebted to many colleagues at Merck Research Laboratories for their input, assistance, and support. At the risk of omitting many, I would like to thank particularly Drs. Joseph F. Heyse, Keith A. Soper, and Daniel J. Holder for their full support and constructive comments; and Drs. Marc Ferrer, Amy S. Espeseth, Berta Strulovici, Raul Lacson, Eric N. Johnson, Ruojing Yang, David Ross, Francesca Santini, Shane D. Marine, Erica M. Stec, Namjin Chung, Weiqing Zhao, Richard A. Klinghoffer, Yaping Liu, Priya Kunapuli, Jayne Chin, Adam Gates, Jenny Tian, Anthony B. Kreamer, Richard R. Peltier, Michael J. Weber, Alexander McCampbell, Louis Locco, Eugen Buehler, David Stone, John Majercak, William J. Ray, and Michele Cleary for their excellent collaboration in RNAi HTS research at Merck. I would like to acknowledge Pamela Peterson and Brenda Holmes for proofreading help; my summer interns Pei-fen Kuan and Xiting Cindy Yang for their assistance; and Drs. Christopher Tong, Xiaoli Shirley Hou, Jason Liao, Yingxue Cathy Liu, Richard Raubertas, and Andy Liaw for their beneficial comments. I would like to thank two anonymous reviewers for their excellent comments and suggestions. I would also like to thank Troop 133 (Chalfont, PA) of Boy Scouts of America for their encouragement when I worked on this book during the 2009 summer camp. Finally and very importantly, I would like to thank my wife, Xiaohua Shu, and my son, Zhaozhi Zek Zhang, for their strong support and great patience during the writing of this book. Without the invaluable input and support of the many professionals I have mentioned and of my family, this book would not have become a reality.

Acronyms and Abbreviations

3'UTR	3' untranslated region
ANOVA	analysis of variance
AUE	approximately unbiased estimate
CI	confidence interval
dsRNA	double-stranded RNA
esiRNA	endoribonuclease-derived siRNA
FDR	false discovery rate
FNDR	false non-discovery rate
FNL	false-negative level
FNR	false-negative rate
FPL	false-positive level
FPR	false-positive rate
HCV	hepatitis C virus
HTS	high-throughput screening
MAD	median absolute deviation
miRNA	microRNA
MLE	maximum likelihood estimate
MM	method of moment
mRNA	messenger RNA
MSE	mean squared error
pnc	proportional noncentral
QC	quality control
RISC	RNA-induced silencing complexes
RNAi	RNA interference
SD	standard deviation
shRNA	short hairpin RNA
siRNA	small interfering RNA

SMCV   standardized mean of a contrast variable
SMLC   standardized mean of a linear combination of random variables
SSMD   strictly standardized mean difference
UMVUE   uniformly minimal variance unbiased estimate
w.r.t.   with respect to

Part I

RNAi HTS and Data Analysis

Introduction to Genome-Scale RNAi Research

1.1 RNAi: An Effective Tool for Elucidating Gene Functions and a New Class of Drugs

RNAi is a mechanism in living cells that helps determine which genes are active and how active they are. It is a naturally occurring pathway for the regulation of gene expression in which small RNA molecules lead to the destruction of *messenger RNA* (mRNA) with complementary nucleotide sequences [48;128]. RNAi has an important role in defending cells against parasitic genes – viruses and transposons – but also in directing development and gene expression in general.

Two types of small RNA molecules are central in the RNAi pathway (Figure 1.1). One is *small interfering RNA* (siRNA), sometimes known as short-interfering RNA or silencing RNA, a class of 20 to 25 nucleotide-long *double-stranded RNA* (dsRNA) molecules [48], and the other is *microRNA* (miRNA), a class of endogenous dsRNA molecules of about 21 to 23 nucleotides in length [89;91;92;128]. Both siRNA and miRNA can bind to other specific RNAs and either increase or decrease their activity, usually by preventing an mRNA from producing a protein.

siRNA. The RNAi pathway is controlled by endoribonuclease-containing complexes known as *RNA-induced silencing complexes* (RISCs) and initiated by an enzyme called *Dicer* in the cell's cytoplasm (Figure 1.1). In the initiation step, the Dicer cleaves long dsRNA molecules into siRNAs. An siRNA assembles into a RISC and unwinds into two single strands. One of the two strands, known as the *guide strand*, is then incorporated into the RISC. Later, the guide strand specifically pairs with a complementary mRNA molecule. This recognition event may produce one of the following two major outcomes: (i) post-transcriptional gene silencing [63;74] (i.e., the gene is not expressed) or (ii) epigenetic changes to a gene affecting the degree to which the gene is transcribed. Post-transcriptional gene silencing occurs when the pairing of the guide strand with mRNA induces cleavage by argonaute, the catalytic component of the RISC complex, in the region homologous to the siRNA.

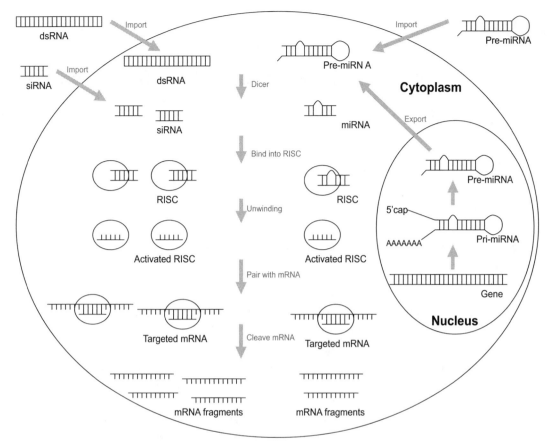

Figure 1.1   The RNA interference pathway. External double-stranded RNAs (dsRNA), siRNA, or pre-miRNA in the RNAi pathway may come from laboratory manipulation, which is the basis for the use of RNAi as an effective tool to knock down targeted genes.

miRNA. The miRNA molecule has a similar function to that of siRNA in the RNAi pathway (Figure 1.1). miRNA is a noncoding RNA. That is, miRNAs are encoded by genes from DNA from which they are transcribed, but miRNAs are not translated into protein; instead, each primary transcript (called a pri-miRNA) is processed into a short, 70-nucleotide stem-loop structure called a pre-miRNA in a cell's nucleus. The stem-loop structure is also called hairpin structure and, subsequently, a pre-miRNA is also called hairpin miRNA precursor [63] or short hairpin RNA (shRNA). A pre-miRNA is exported into cytoplasm and is then identified and cleaved by Dicers into a functional miRNA. Mature miRNAs are structurally similar to siRNAs. However, mature miRNA molecules are partially complementary to one or more mRNA molecules. The main function of miRNA is thought to be down-regulation of the translation of mRNA to protein.

In the RNAi pathway, the dsRNA may come from infection by a virus with an RNA genome or from laboratory manipulations. Therefore, this pathway can be co-opted by experimentally introducing synthetic dsRNAs designed to target specific mRNAs, thus knocking down the expression of the protein of interest [44;63;64]. The development of algorithms for siRNA design that produce potent and selective knockdown of targeted genes has led to a great deal of interest in using siRNA to elucidate gene function and identify novel targets for drug discovery. In medical research, in addition to drug target identification and validation, RNAi can also be harnessed to develop a whole new class of potential therapeutic agents [98]. In fact, RNAi is seen as the third class of drugs, after small molecules and proteins [24]. The importance of RNAi was further recognized when the Nobel Prize in Medicine and Physiology was awarded to A. Fire and C.C. Mello in 2006 for their work [48] on RNA interference in *Caenorhabditis elegans*, which they published in 1998. Galun [51] even parallels the article by Fire et al. [48] on RNAi to that of Watson and Crick [156] on the double helix of DNA.

1.2 High-Throughput Screening: A Vital Technology in Drug Discovery

High-throughput technologies such as microarrays, whole-genome single-nucleotide polymorphism chips, and *high-throughput screening* (HTS) play a central role in current molecular biological research and drug discovery. The ability of high-throughput technologies to simultaneously interrogate thousands of genes/compounds has led to important advances in solving a wide range of biological problems, including the identification of previously unknown genes involved in a biological pathway and the subsequent unveiling of new insights into developmental processes and pharmacogenomic responses, the evolution of gene regulation, and the discovery of new drug targets [18;54;100]. Likewise, RNAi can be utilized on a genome-wide scale via HTS technology, which allows thousands of siRNAs to be tested simultaneously to identify previously unknown genes involved in a biological pathway [8;16;64;96;138;158;176;180].

HTS technology uses automation (including robotics, data processing, and control software, liquid handling devices, and sensitive detectors) to run an assay screen against a library of candidate compounds or siRNAs. An assay is a test for specific biochemical activity such as the inhibition or stimulation of a biochemical or biological mechanism. The biochemical activity can be represented by measured responses such as the reflectivity of polarized light shined on cells or the intensity of emission from labeled particles. A typical compound HTS-screening library contains more than 100,000 small molecules. A genome-scale siRNA library may contain 60,000 or more siRNAs that are pooled to target about 25,000 genes. Usually, three siRNAs targeting the same gene are pooled together. Hence, using HTS, one can rapidly identify active compounds, antibodies, genes, or effective siRNAs that modulate a

Figure 1.2   Two types of microtiter plates commonly used in high-throughput screens. Left: a 384-well plate. Right: a 1,536-well plate.

particular biomolecular pathway and thus can discover the interaction or role of a particular biochemical process in biology.

The key testing vessel of HTS is a small container with a grid of small, open divots, called wells. This container is called a microtiter plate or microplate of about 5 inches long and $3\frac{1}{8}$ inches wide and is usually disposable and made of plastic (Figure 1.2). The microplates for HTS generally have 384, 1,536, or 3,456 wells, although they may have 96 wells in some experiments. Most of the wells contain test compounds (i.e., small molecules) or siRNAs, with one compound or siRNA per well, although some of the wells contain controls to indicate the quality of the assays. The test compounds or siRNAs are also called sample compounds or sample siRNAs. A screening facility usually has a library of source microtiter plates holding the compounds or siRNAs that have been carefully chosen, arranged, and cataloged. The source plates are also called stock plates, and they are either created by the lab or obtained from a commercial source. The source plates themselves are not directly used in experiments. During experiments, copies of selected source plates are created by pipetting a small amount of liquid (often measured in nanoliters) from the wells of a source plate to the corresponding wells of a completely empty plate. The copied plates that are actually used in the experiments are called *assay plates*.

To conduct an HTS experiment, the researchers fill each well of an assay plate with some biological matter, such as protein, cells, or an animal embryo, and incubate them for a certain time so that the biological matter can absorb, bind to, or have other reactions with the compounds/siRNAs in the wells. Then the response representing the biochemical reaction (e.g., the intensity of light emitted by labeled particles) is measured across the wells, usually by an automated machine. The machine outputs the result as a grid of numeric values, with each number mapping to the value obtained from a single well of an assay plate in an experiment. A high-capacity machine can measure dozens of assay plates in a few minutes, generating very quickly thousands of experimental data points. One of the most important features in HTS technology is automation, which relies on robotics and high-speed computers. Typically, an integrated HTS system consisting of one or more robots has the ability to transport assay plates from station to station for sample and reagent addition, mixing, incubation, and readout. Therefore, an HTS system can usually simultaneously prepare, incubate, and output many plates, subsequently testing a large number of compounds/siRNAs (e.g., up to 100,000 compounds per day) and generating a huge amount of data. The term *ultra high-throughput screening* (uHTS) has been created to refer to an HTS facility that can screen in excess of 100,000 compounds a day on a routine basis.

1.3 Genome-Scale RNAi Screens

Genome-scale RNAi screens can be conducted in different organisms. Three that have been intensively studied are *C. elegans* (a small roundworm), *Drosophila* (fruit flies), and human cells.

Human genome-scale RNAi screens are currently conducted in human cells, including stem cells, and a variety of immortal cell lines [41]. Long dsRNAs activate interferon responses, which leads to apoptosis (cell death) in somatic cells. Short dsRNAs do not activate the interferon response; thus RNAi in human cells must use short dsRNA of less than 30 nucleotides such as synthetic siRNA, vector-expressed *short-hairpin RNA* (shRNA, and *endoribonuclease-derived siRNA* (esiRNA). Even shorter dsRNAs may be needed: a recent study [85] found that dsRNAs of just 21 nucleotides long triggered an immune response through toll-like receptor three (TLR-3), which ultimately inhibited angiogenesis, and that simply shortening the siRNA to fewer than 18 nucleotides seemed to eliminate TLR-3 recognition.

RNAi screens in human cells can use a diverse set of phenotypic measurements such as homogenous cell viability, alternations in reporter-gene expression, high-content readouts using automated microscopy, and immunofluorescence signal from a highly specific antibody. In situations where the focus is on a single measured response, the design and analysis should be similar regardless of differing types of phenotypic measurements. In other cases, we may consider multiple measurements simultaneously in an experiment, especially for high-content readout. More details about different analyses will be provided in Chapter 5.

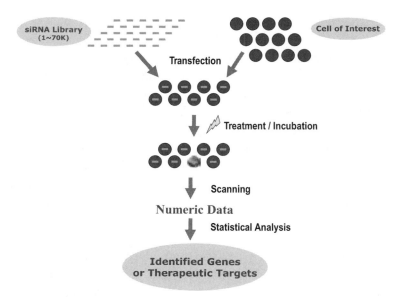

Figure 1.3   Procedure of genome-scale RNAi screens.

The HTS technology applied to RNAi has made it feasible to use cell-based assays to query every gene in the genome for its potential function in a given biological process of a cell. The general procedure for cell-based RNAi screens is demonstrated in Figure 1.3. The first step is to choose an RNAi library and a robust and stable type of cell, which can be a stem cell, primary cell, established cell line, or engineered cell line. Primary cells have limited passaging capacity (thus siRNAs are hard to get into the cells), whereas established cell lines such as Hela and HEK are capable of indefinite passaging under proper cell culture conditions. Engineered cell lines are modified to over-express or underexpress native, tagged, or engineered proteins.

Transfection. The delivery of siRNAs in a library into cells is called *transfection* of siRNAs into the cells in an RNAi screen. For transfection, we need to identify relevant siRNA controls and HTS-compatible transfection methods that give optimal gene knockdown and cell viability. The transfection can be conducted using either suspension-mode electroporation or lipid reagents. Suspension-mode electroporation is good for cells that are difficult to transfect but currently is limited to the 96-well format. Lipid reagents are easily scalable and mostly highly efficient with varying degrees of cytotoxicity and stability. In addition to choosing the best transfection method, we need to determine cell density (i.e., number of cells per well), time of lipid/microRNA complex formation, assay incubation time post-transfection, transfection efficiency (i.e., the performance of assay with respect to positive and negative control siRNAs), detection reagent stability at the working concentration, effect of cell passage number, and so on.

There are various types of RNAi-detection assay technologies. Two commonly used types are well-based assays using photomultiplier tube– or charge-coupled device–based cameras and cell imaging based on fluorescent labeling of macro-molecule of interest. To minimize the need to continually change platform instru-mentation for new assays, RNAi assays can be broken into portions, with one robotic platform handling transfection, a second handling detection, and, possibly, a third carrying out a high-content read, which allows each robot to specialize in a particular aspect of the RNAi HTS process.

1.4 An Example of Genome-Scale RNAi Research

In genome-scale RNAi screens, a primary goal is to select siRNAs with a desired effect size. The siRNA effect is represented by the magnitude of difference between the intensity of an siRNA and that of a negative reference in RNAi HTS experiments [183]. For screens using the common platform of 384-well plates, limitations of experimental time and cost usually do not allow a single experiment to have more than two hundred 384-well plates, whereas two hundred 384-well plates is usually the minimal requirement for conducting a genome-wide screen with replicates (i.e., each siRNA is measured multiple times). Therefore, currently, a typical RNAi HTS project starts with a first screen (called *primary screen*) of single or pooled siRNAs targeting about 20,000 genes, most of which have no replicate. The single or pooled siRNAs identified (called hits) in the primary screen are further investigated using one or more secondary screens (called *confirmatory screens*) in which each siRNA or pool has replicates. A typical primary screen has fifty to one hundred fifty 384-well plates, and a typical confirmatory screen has three to twenty 384-well plates.

For example, a genome-scale RNAi project for hepatitis C virus (HCV) started with a primary screen, in which a total of about 22,000 siRNA pools were tested across 97 plates [180]. The experiment was designed to identify host factors associated with HCV replication using the HCV replicon assay system described in Zuck et al. [185]. The negative control used in the experiment was a nonsilencing siRNA. Two positive control siRNAs were used: (i) a very strong one that targeted the HCV replicon [121] and (ii) a weaker one that targeted hVAP33 [61;97].

Following the primary RNAi HTS experiment and using the methods described in Zhang et al. [161;162;174;180], a total of 640 siRNAs were identified as hits. These siRNAs were transfected into 384-well plates with controls set up as in the primary experiment. HuH-7 cells containing an HCV genotype 1b replicon were transfected with these 640 siRNAs as described for the primary screen, but the transfections were carried out in triplicate of every source plate for improved statistical robustness in a confirmatory screen. In addition, a second confirmation screen was carried out using HuH-7 cells expressing the genotype BK-2b HCV replicon. The replicon-containing cells were transfected in triplicate and assayed in the same manner as the HuH-7 genotype 1b HCV replicon containing cells in the primary screen and in the first confirmatory screen [173].

1.5 Challenges in Genome-Scale RNAi Research

Genome-scale RNAi screens have two major advantages over classical genetic screens: (i) the sequences of all identified genes are immediately known, and (ii) lethal mutations are easier to identify because it is unnecessary to recover mutants [16]. Classical genetic screens for elucidating gene function largely rely on the recovery of lethal mutation, which is usually time-consuming and may be difficult. Now RNAi screens directly measure the knock-down impact of siRNAs on their targeted genes, thus making it unnecessary to recover mutants as in classical genetic screens. Thus genome-scale RNAi screens have great promise for elucidating gene function and for discovering new drug targets in our post-genome era. Meanwhile, as a technology that is still under development, genome-scale RNAi screens face many challenges. Four key challenges are (i) controlling optimal experimental time, (ii) identifying moderate or weak hits, (iii) reducing off-target effects, and (iv) gleaning biologically relevant information from a large body of data.

Experimental times. Compared with small-molecule HTS experiments, RNAi HTS experiments have more challenges in controlling the optimal experimental time so that all the potent siRNAs are measured in their effective peaks. RNAi targets mRNA and depletes it from the cell. However, the levels of mRNA vary across a wide dynamic range; consequently, the depletion of different mRNAs may take different lengths of time. Furthermore, after the depletion of mRNAs in the cell, the residual protein can remain for an extended time, which is protein dependent. As a result, the knockdown of genes by RNAi reagents has a wide temporal range (e.g., within 12–120 hours after transfection) [62]. When a large number of siRNAs are used in a single assay as in a genome-scale RNAi screen, the onset of action of potent siRNAs can occur at different times, and there is no time point in the assay that is optimal for all the potent siRNAs. If a potent siRNA is not measured at or nearly at its effective peak, it may act like an impotent siRNA during measurement and may not be identified as a hit in the assay, which then leads to a higher false-negative rate in RNAi screens than in small-molecule screens.

Identifying hits. A unique feature of RNAi is that its effect on genes is to knock down, not knock out completely. Compared with the effect of knockout in classical genetic screens, the size of effect on a measured phenotype is moderate or weak for many potent siRNAs. Furthermore, in some cases, the siRNAs with moderate effects are more biologically relevant [173]. Although many of the siRNAs might affect the outcome of the assay, in most cases, a small percentage of the effective siRNAs truly target genes involved in the phenotype under investigation. That is, the true-positive hits are buried in a large body of data containing substantial noise. siRNAs with moderate or weak effects are more likely to be buried among siRNAs with extremely weak or no effects. In other words, a nonhit tends to behave more like a hit in RNAi HTS experiments. In contrast, the effects of potent small molecules are usually strong. Therefore, RNAi HTS experiments tend to have a higher false-positive rate than both classical genetic screening experiments and small-molecule HTS experiments.

Off-target effects. When using RNAi as a gene-silencing tool, we want an RNAi reagent to specifically knock down a target gene but not to interfere with other genes. However, an siRNA can silence not only the target gene, but also other genes with similar sequences. The silencing effect of an siRNA on nontarget genes is called an *off-target effect*. Off-target effects can produce false positives, leading to misleading results and erroneous conclusions about the genes that are involved in a biological pathway, when RNAi experiments are used to elucidate gene functions [78–80].

Detecting biologically relevant data. One of the major advantages of HTS technologies is their ability to simultaneously interrogate thousands of genes/compounds. With the ability to generate large amounts of data per experiment, HTS technologies have led to an explosion in the rate of data compiled in recent years [9;70]. Consequently, one of the most fundamental challenges of HTS biotechnologies is to glean biological significance from large volumes of data [16;25;38; 43;70;82;136;161;176;180]. There are many analytic questions to be solved. For example, systematic errors (e.g., row and/or column effects) and outliers are not uncommon in HTS experiments. How should we address them? For quality control (QC) in genome-scale RNAi experiments, signal-to-background ratio, signal-to-noise ratio, signal window, assay variability ratio, and Z-factor have been adopted to evaluate data quality [17;39;77;99;116;123;148;150;159;180;185]. How well do these QC metrics work? For hit selection, z-score and t-statistic are commonly used. Do these methods work well? If not, can we develop better analytic methods? Should we perform analyses on a plate-by-plate basis (called plate-wise) or on all the plates in an experiment (called experiment-wise)? How should we address multiplicity issues of hit selection in genome-scale RNAi screens?

To face all these challenges, we must adopt appropriate experimental designs and suitable analytic methods so that we can obtain optimal results in genome-scale RNAi research. For example, the z-score and t-statistic are both based on testing the null hypothesis of exactly no effects on average. However, owing to the network of gene interactions, many genes may have some degree of impact on a measured biochemical response [167;178]. Better analytic metrics are required for assessing the size of siRNA effects rather than testing the null hypothesis of no effect on average. I have developed a statistical parameter, *strictly standardized mean difference* (SSMD), for effectively measuring the size of siRNA effects [161;162;165]. On the basis of SSMD, I have also proposed an error-control method for maintaining a balanced control of both false-positive and false-negative rates [161;174;175;178]. In addition, we need to adopt effective sequence designs, such as the design of multiple individual siRNAs per gene and siRNA pooling to address off-target effects [16]. We also need better plate designs to account for positional effects in RNAi screens [166;173]. In the following chapters, I present the use of experimental designs and analytic methods to optimize genome-scale RNAi research. Specifically, I discuss experiment designs in Chapter 2, data display and normalization in Chapter 3, quality control in Chapter 4, hit selection for screens without replicates in Chapter 5, and hit selection for screens with replicates in Chapter 6.

2

Experimental Designs

As illustrated in Figure 1.3 of Chapter 1, the basic procedure of a cell-based genome-scale RNAi screen includes selection of an RNAi library to be screened, choice of human cells, transfection of siRNA into cells, treatment or incubation, detection, and statistical analysis. Following this procedure, the success of a cell-based genome-scale RNAi screen in this procedure relies on the design of the following elements: siRNA, control, plate, sample size, and methods for optimizing siRNA delivery and transfection efficiency. All these designs are explored in this chapter.

Beginning with this chapter, most of the work discussed in this book focuses on RNAi screens using siRNA; however, the designs and methods described in the following chapters are also applicable to RNAi screens performed in microplates with other silencing reagents, including shRNA, esiRNA, and dsRNA, which are described in Chapter 1.

2.1 siRNA Designs

The first step in starting a genome-scale RNAi screen is to choose an RNAi library. One important criterion for an siRNA library is well-designed siRNAs. siRNA design is critical for both successful gene knockdown and on-target hit selection. The quality of siRNA design is gauged mainly by the potency and specificity of the siRNAs in a library. siRNA design is sometimes performed by commercial siRNA vendors. In other cases, researchers also must get involved in siRNA design by either doing it alone or cooperating with the vendors.

RNAi is the phenomenon in which double-stranded RNA knocks down a gene in a sequence-specific manner. Thus one should expect that the introduction of an siRNA can produce a specific effect mediated by the gene that this siRNA targets. The potency of an siRNA is represented by the total amount of the phenotypic effect related to the introduction of the siRNA into the cells or an organism. The design of the siRNA sequence is crucial for finding the siRNAs that produce a high phenotypic effect. The specificity of an siRNA is represented by its on-target gene knockdown; namely, the amount of phenotypic effect that is caused by its knockdown on its target gene. However, the introduction of siRNA can result in nonspecific phenotypic effects. These nonspecific effects appear to have three separate origins [23]: (i) transfection agent-mediated response [47], (ii) interferon response [69;81;85], and (iii) siRNA-mediated off-target effects [12;46;78–80;94;130;133;137].

Transfection is the delivery of siRNAs in a library into the cells in an RNAi screen. RNA cannot penetrate cellular membranes. Thus the delivery of siRNAs to target cells requires specific methods, such as viral delivery, the use of liposomes or nanoparticles, bacterial delivery, and chemical modification of siRNA to improve stability. Currently, for cell-based genome-scale RNAi screens, an RNAi library is constructed using suspension-mode electroporation, lipid reagents, or viral backbones.

Transfection agents such as liposomes that are used to carry siRNA into cells may induce broad changes in gene expression profiles [47]. Long dsRNAs can activate interferon response in a cell-type specific manner [69;81;85]. In general, short dsRNA of less than 30 nucleotides can evade the interferon response, but some shorter dsRNAs have been observed to trigger an immune response. The nonspecific effects caused by transfection agent–mediated response can be eliminated by adopting or developing better delivery methods (see Section 2.4) or by optimizing lipid concentrations/compositions. The interferon response can be eliminated by adopting stringent siRNA design filters [81;85].

Off-target effects. Off-target siRNA-mediated effects offer a more serious challenge in the application of RNAi in biomedical research and drug discovery and development [23]. In this case, unintended mRNA targets with sequence homology to the RNAi oligonucleotide are knocked down in addition to, or instead of, the intended target gene. Off-target gene knockdown may induce measurable phenotypes [46;94]. False positives generated by off-targets during phenotypic screens can result in false leads, the use of resources to explore nonproductive research tracks, and even serious safety issues in therapeutic agents.

Off-target effects can be addressed in both the experimental design stage and the data analysis stage. In the experimental design stage, the focus is on the following major approaches to siRNA design: siRNA pooling [23], siRNA modification [79], siRNA construct design using 3′ untranslated region (3′UTR) seed match [12;78;94], and design of multiple siRNAs per gene. In siRNA pooling, several single siRNA sequences targeting the same gene are arranged in each well of a plate in an RNAi screen. The rationale behind siRNA pooling is that off-targets are concentration dependent. A pool consisting of multiple siRNAs that target the same gene can maintain a desired size of on-target gene knockdown while inducing only a fraction of the off-target effects that one or more of the pooled siRNAs might have. The siRNA modification approach is to use chemical modification to reduce off-target effects; for example, the differential addition of 2′-O-methyl moieties to both the sense and antisense strands of the siRNA [79]. 3′UTR seed match is defined as the presence of one or more perfect matches between the 3′UTR of a gene and the hexamer or heptamer seed region (positions 2–7 or 2–8) of the antisense strand of the siRNA, which can increase the occurrence of off-target effects [12;78;94]. Utilizing this observation, the siRNA sequences that have a 3′UTR seed match with unintended genes during the siRNA construction stage can be replaced or removed.

Deconvolution screens. A more accessible solution to the problem of off-target effects that has been adopted by most researchers and companies today is to design and conduct an experiment in which multiple siRNAs with different sequences are tested against a target gene to increase the level of confidence in positive hits [14;40]. These types of experiments, usually called *deconvolution screens*, are usually conducted after reducing the selected hits to a limited number of approximately 100 to 800 from primary and/or confirmatory screens. In a deconvolution screen, 3 to 7 siRNAs per gene are typically measured, and the collective activity of multiple siRNAs targeting the same gene is analyzed. More details about the analysis of deconvolution screens are discussed in Chapters 6 and 8.

2.2 Control Designs

Effective positive and negative controls should be selected to develop the RNAi screen assay to achieve high signaling with the positive controls and low noise with the negative ones [16]. Negative control siRNAs should be designed to have no specific effects on any characterized genes in the experiment [173]. Whenever possible, positive controls should encompass a range of strengths to develop an assay that can identify both weak and strong hits [16;173]. The design of effective biological controls is critical for conducting meaningful experiments.

2.2.1 Design of Negative Controls

Negative controls are important to monitor data variability and systematic experimental errors. They can serve as a negative reference for hit selection, especially in confirmatory screens. Negative control siRNAs are usually purchased and are designed to be *nonsilencing*, meaning that they do not target any characterized genes. Negative control siRNAs can be those with no homology to any known mammalian gene or with homology to a gene that is not present in the cells under study.

Unlike small-molecule screens, which can utilize vehicle-only wells as negative controls, RNAi screens do not have such universal negative controls. This is because RNAi screens usually involve complex delivery vehicles that can also have biological effects, and even nontargeting siRNA controls may exhibit off-target effects in some cell lines [78]. It is essential to test multiple negative control siRNAs to identify one that has the smallest effect relative to mock transfection (typically, introduction of transfection reagents in the absence of RNA).

2.2.2 Design of Positive Controls

Positive controls are used to monitor the quality of data produced during the screen and aid in the selection of physiologically relevant hits. Positive controls are typically siRNAs targeting genes with a known association to the biological process being studied and resulting in a particular phenotypic effect. They may also be siRNAs that are known to result in high gene knockdown.

A challenge in identifying positive control siRNAs is that siRNAs are typically more effective if they target mRNAs present in limited amounts for the biological process being assayed; as such, even proteins with well-characterized associations with a biological process may not make effective siRNA controls. For example, the lens epithelium-derived growth factor (LEDGF) protein is critical for human immunodeficiency virus (HIV) integrase function, but only a small fraction of the LEDGF proteins present in the cell is required for this function [27;95;97]. To see an effect of siRNA-mediated LEDGF knockdown on HIV infection, cells must be subjected to intensified RNA interference, which is impractical for HTS. Therefore, LEDGF is an invalid positive control for any siRNA HTS for HIV host factors, despite its biological relevance. Positive controls may provide more of a challenge than negative controls if the biology of the process being studied is not well understood. Ideally, more than one positive control will be used, as in, for example, the HCV screens described in Section 1.4 of Chapter 1, in which two positive control siRNAs were used to inhibit HCV replication [180].

An additional consideration is that the positive controls should be as effective as the hits in the screen are expected to be. An example of this is shown in the HCV screens described in Section 1.4 of Chapter 1. The very strong positive control siRNA targeted the HCV replicon and was uniformly effective at knocking down 90% to 95% of HCV replication. In contrast, the weaker positive control targeted hVAP33, a host factor known to participate in HCV replication. In this instance, the weaker positive control was more instructive than the very strong positive control, because the effectiveness of this control was more similar to HTS hits than the siRNA targeting HCV itself. Another example of fairly strong control is epidermal growth factor receptor in a mucin screen [174].

2.3 Plate Designs

The arrangement of control and sample wells in a plate is called *plate design*. The existence of systematic errors of measurement is not uncommon in HTS experiments [20;58;99;143]. A good plate design helps to identify systematic errors (especially those linked with well position) and determine what normalization should be used to adjust the data so that the impact of systematic errors on both QC and hit selection can be removed or reduced. Here, various plate designs are compared.

First, we need to choose a plate format. The current formats are 96-well, 384-well, and 1,536-well plates. Currently, the 384-well plate is the commonly used format for a genome-scale RNAi screen in which approximately 25,000 siRNAs are screened, although the use of the 1,536-well plate format is increasing [166]. The disadvantage of the 96-well plate format is its low capacity for containing siRNAs in a plate, thus requiring too many plates for a genome-scale RNAi screen. The advantage of the 1,536-well plate format is its better capacity for including more siRNAs and more replicates in a genome-scale screen. Currently, the use of 96-well or 384-well plate formats barely allows the arrangement of replicates for each siRNA in a genome-scale primary screen. By contrast, the use of the 1,536-well plate format can allow three or four replicates per siRNA in a primary screen. The disadvantage of the 1,536-well plate format, however, is that the intensity of a well with a strong effect is more likely to affect the measured intensity of its neighbor wells because the size of a well in this format is much smaller (i.e., only one quarter of the size of a well in a 384-well plate [Figure 1.2]).

For a primary screen, sample siRNAs should be randomly arranged across the plates, as well as across the well positions in each plate. This is because it is difficult to distinguish enriched plates (or enriched areas within a plate) with plate-to-plate variation (or systematic positional effects) in the late process of hit selection. Systematic errors such as plate-to-plate variation and systematic spatial effects commonly exist in RNAi screens.

2.3.1 Construction of Plate Designs

Because genome-scale RNAi screens are most often conducted in 384-well plates, we focus here on the design of wells in such plates. A typical format for current plate designs involves arranging multiple types of controls in the four edge columns and arranging sample siRNAs in the remaining columns (e.g., designs A and B in Figure 2.1). Design A (or similar designs) is commonly used in current RNAi HTS experiments because it is more convenient for laboratory staff to arrange one control in a column consecutively. In design B, the negative control wells are arranged in both edges and occupy both top and bottom rows. The positive control wells are arranged similarly. Thus design B is more balanced than design A. Both designs A and B are subject to the restriction in which control wells can only be arranged in the four edge columns. The restriction of designs A and B is that they can only be used to display or adjust a few types of positional effects (as is shown in Figures 2.2, 2.3, 2.4, 2.5). Designs with controls arranged in the middle of a plate (such as in designs C–F) are needed for effectively displaying and adjusting various commonly existing positional effects.

Now that robots can readily arrange controls anywhere in a plate, it is possible to adopt designs C through F (Figure 2.1). The arrangement of the negative control in design C is the same as in design D, which can effectively display various patterns of systematic errors, especially for those with a sharp change in the edge row/column. Design E is a variant of design C or D that allows for one more column to accommodate other controls. Similar variants can be obtained to allow for four edge columns to accommodate other controls. Design F may not work as effectively as designs C through E, especially when there is a sharp change in the edge; however, fewer wells are needed for the negative control (16 in design F as compared with 24 in

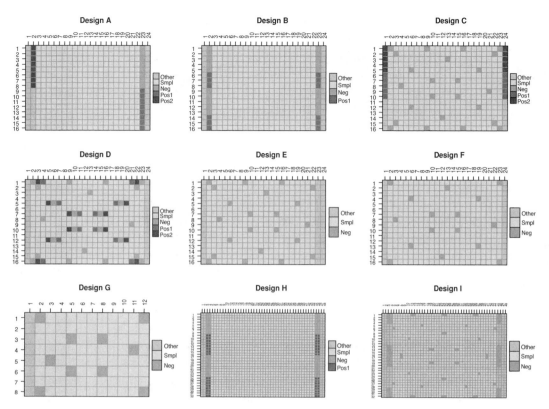

Figure 2.1   (See color insert following page 110.) Plate designs in a 384-well (designs A–F), 96-well (design G), or 1,536-well plate (designs H and I). Designs A, B, and H are for situations in which controls can only be arranged in edge columns; designs C through G and I are for situations in which controls are allowed to be arranged anywhere in a plate. The colors represent types of wells, as shown in the legend to the right of each panel: green = negative control; red = first positive control; purple = second positive control; yellow = sample siRNAs; gray = other controls. *Source:* From Zhang [166].

designs C and D). For design F, positive controls can be arranged in the edge, similar to design C, or in the middle, similar to design D. Among these designs, design D has the highest capacity for displaying various positional effects; however, it requires positive controls to be arranged in the middle of a plate, which can be difficult and time-consuming for many researchers. The major difference among designs C, E, and F is their different capacity for accommodating controls.

The plate designs A through F in Figure 2.1 are constructed for experiments with a 384-well plate. Similar plate designs and guidelines can be constructed for experiments with a 96-well or 1,536-well plate. Design G is a design for a 96-well plate, and designs H and I are for a 1,536-well plate. Design H is subject to the restriction by which the control wells can be arranged only in the eight edge columns. In a

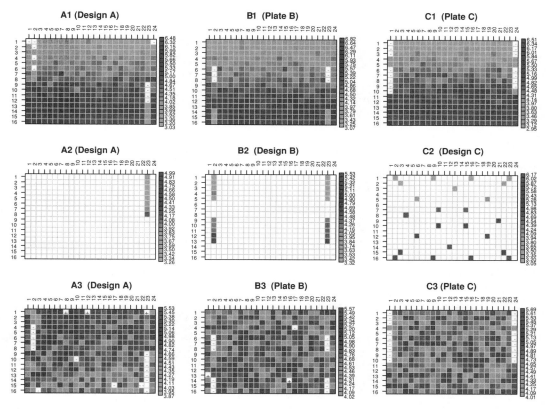

Figure 2.2   (See color insert following page 110.) The capacity of three typical plate designs, designs A, B, and C (shown in Figure 2.1), in identifying and adjusting for linear row effects. Panels A1, B1, and C1 show the measured intensities (in log10 scale) in all wells in a plate from three experiments that have plate designs A, B, and C, respectively. Panels A2, B2, and C2 display intensities of the negative control. Panels A3, B3, and C3 display the data adjusted using the negative control wells. In each panel, a red + (or a green −) denotes an outlier in up-regulated (or down-regulated) direction based on sample wells.

1,536-well plate, the intensity of a well with a strong effect is more likely to affect the measured intensity of its neighbor wells; thus in design I, we arrange many negative control wells in the 3rd, 4th, 45th, and 46th columns to establish buffering borders. When the impact of edge wells on the neighbor wells is strong, the negative control wells in these four columns are not used for smoothing, but are used instead for the comparison with positive controls in the edge columns. For experiments conducted in 1,536-well plates, only one primary screen with replicates may be needed, with no follow-up confirmatory screens. In such a case, to adopt design I or similar designs, it may be necessary to work with vendors of genomic libraries to arrange a control such as luciferase in the green wells during the process of generating a genomic library.

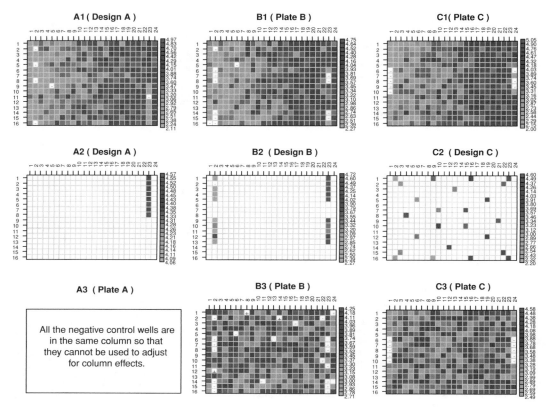

Figure 2.3 (See color insert following page 110.) The capacity of three typical plate designs, designs A, B, and C (shown in Figure 2.1), in identifying and adjusting for linear column effects. Panels A1, B1, and C1 show the measured intensities (in log10 scale) in all wells in a plate from three experiments that have plate designs A, B, and C, respectively. Panels A2, B2, and C2 display intensities of the negative control. Panels A3, B3, and C3 display the data adjusted using the negative control wells. In each panel, a red + (or a green –) denotes an outlier in up-regulated (or down-regulated) direction based on sample wells.

2.3.2 Capacity of Plate Designs in Identifying and Adjusting Spatial Effects

Systematic spatial or positional effects commonly exist in genome-scale RNAi screens. In many screens, especially in confirmatory screens, it is hard to determine whether systematic spatial effects are caused by systematic experimental errors or by the location of true hits if an effective plate design has not been adopted. Effective plate designs can greatly help to identify and adjust systematic experimental errors in RNAi screens.

Systematic spatial effects occur in a variety of patterns. Four common patterns are (i) linear row effect, in which measured raw or transformed values have a linear relationship to row numbers (illustrated in Figure 2.2A1, B1, and C1); (ii) linear column effect, in which measured raw or transformed values have a linear relationship to column numbers (illustrated in Figure 2.3A1, B1, and C1); (iii) linear row and

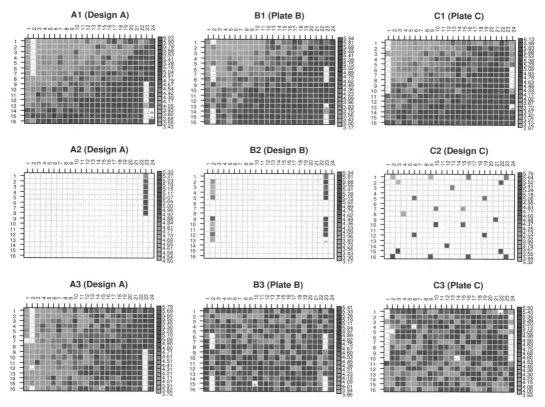

Figure 2.4   (See color insert following page 110.) The capacity of three typical plate designs, designs A, B, and C (shown in Figure 2.1), in identifying and adjusting for linear row and column effects. Panels A1, B1, and C1 show the measured intensities (in log10 scale) in all wells in a plate from three experiments that have plate designs A, B, and C, respectively. Panels A2, B2, and C2 display intensities of the negative control. Panels A3, B3, and C3 display the data adjusted using the negative control wells. In each panel, a red + (or a green −) denotes an outlier in up-regulated (or down-regulated) direction based on sample wells.

column effect, in which measured raw or transformed values have a linear relationship to both row and column numbers (illustrated in Figure 2.4A1, B1, and C1); and (iv) bowl-shaped spatial effect, in which measured raw or transformed values have a bowled relationship (either concave or convex) to well positions (illustrated in Figure 2.5A1, B1, and C1). The spatial effects can be revealed using plate image plot and plate-well series plot, which is described in Chapter 3. Here I show the capacity of three typical plate designs, designs A, B, and C shown in Figure 2.1, in identifying and adjusting for the four common patterns of systematic spatial effects.

Row effects. In design A, because the negative control wells only occupy the upper eight rows in a column, they cannot reveal the spatial effect in the bottom eight rows (Figure 2.2A2) and may not be used successfully to adjust for the row effects in the bottom eight rows (Figure 2.2A3). In design B, the negative control wells only

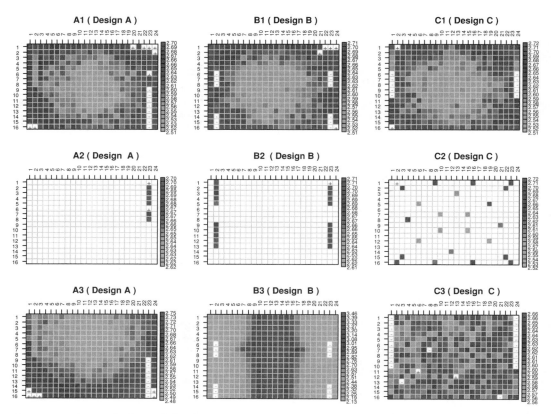

Figure 2.5   (See color insert following page 110.) The capacity of three typical plate designs, designs A, B, and C (shown in Figure 2.1), in identifying and adjusting for bowl-shaped spatial effects. Panels A1, B1, and C1 show the measured intensities (in log10 scale) in all wells in a plate from three experiments that have plate designs A, B, and C, respectively. Panels A2, B2, and C2 display intensities of the negative control. Panels A3, B3, and C3 display the data adjusted using the negative control wells. In each panel, a red + (or a green –) denotes an outlier in up-regulated (or down-regulated) direction based on sample wells. *Source:* From Zhang [166].

occupy five upper rows and five bottom rows, both in two columns; thus they can reveal the row effect (Figure 2.2B2) and can be used to adjust for the row effects in all rows (Figure 2.2B3). In design C, the negative control wells are balanced among 16 rows; thus they can reveal the row effects (Figure 2.2C2) and can be used to adjust for the row effects in all rows better than design B (Figure 2.2C3).

Column effects. In design A, because all the negative control wells are in the same column, they cannot reveal the column effects (Figure 2.3A2) and cannot be used to adjust for the column effects (Figure 2.3A3). In design B, the negative control wells are located in both a left column and a right column; thus they can reveal the column effects (Figure 2.3B2) and can be used successfully to adjust for the column effects (Figure 2.3B3). In design C, the negative control wells are balanced among

14 columns; thus they can reveal the column effects (Figure 2.3C2) and can adjust for the column effects in all rows better than design B (Figure 2.3C3).

Similarly, as illustrated in Figure 2.4, design A can neither reveal nor be used to adjust for linear row and column effects, whereas both designs B and C can effectively reveal the linear row and column effects and be used to adjust for these effects.

Bowl-shaped spatial effects. In designs A and B in Figure 2.5, no negative control wells are located in the middle of the plate; therefore, neither design can reveal the spatial effects or be used to adjust for bowl-shaped effects (panels A2 and B2). The negative control wells in design C are balanced in the middle; thus they can reveal and be used to adjust for the bowl-shaped spatial effects (Figure 2.5C2 and C3).

2.3.3 Guidelines for Adopting Plate Designs in HTS Experiments

In primary screens, we may assume that the majority of sample siRNA wells have no effects. Thus it is reasonable to use the majority of sample wells to adjust for systematic errors of measurement. However, even if we do so, plate designs with controls arranged only in edge columns cannot identify the plates with enriched hits. Given these facts, I suggest the following guideline for choosing plate designs in primary screens: if possible, adopt one of designs C through F and their variants, especially in a situation in which sample siRNAs are not randomly arranged in the screen; otherwise, adopt design B or a similar design. To adopt designs C through F or their variants in primary screens, it may be necessary to work with vendors of genomic libraries so that a control such as luciferase or polo-like kinase-1 can be arranged in the green wells of a plate during the process of generating a genomic library.

In confirmatory screens, it is infeasible to use sample wells to adjust for systematic errors because the sample siRNAs are pre-selected to have inhibition or activation effects. Given this fact, I suggest the following guideline for choosing plate designs in confirmatory screens: adopt one of designs C through F and their variants whenever possible; adopt design B or a similar design and avoid design A or a similar design if the negative control cannot be arranged in the middle of a plate.

Use of designs C through F or their variants depends on (i) the ability to arrange positive control wells in the middle of a plate and (ii) the tradeoff between the ability of the chosen design to display systematic errors and the number of negative control wells. If the positive control cannot be arranged in the middle of a plate or if the arrangement may greatly affect the measured intensities of neighbor wells, choose a design with positive controls only in the edge columns, such as design C; otherwise, choose a design with positive controls in the middle of a plate, such as design D. If more wells are needed in the edge for other types of controls, choose design E or a similar design. If more wells are needed for sample siRNAs, choose design F or one of their variants.

The preceding plate designs and guidelines are intended for experiments with a 384-well plate. Similar plate designs and guidelines can be constructed for experiments with a 96-well or 1,536-well plate, such as design G for a 96-well plate and designs H and I for a 1,536-well plate.

The above guidelines for adopting plate designs are important and necessary in designing a genome-scale RNAi screen. In Chapter 3, we will see how the absence of an effective plate design can produce misleading results of hit selection, and in Chapter 4, we will see that effective plate design is needed for quality control, whether in the primary/confirmatory screen, in which we use analytic metrics to measure the differentiation between a positive control and a negative control, or in the confirmatory screen, in which we use the correlation analysis for two replicate plates from the same source plate. The adoption of effective plate designs can help to reduce or eliminate such misleading results in both quality control and hit selection [166;173].

2.4 Designs of siRNA Delivery and Optimization of Transfection

Positive control siRNAs should be designed in every experiment to help monitor transfection efficiency. One commonly used positive control for transfection in cell-based RNAi screens is PLK1. Meanwhile, negative control siRNAs such as luciferase should be designed to monitor cytotoxicity resulting from the siRNA delivery method. To do so, the viable cell numbers in cultures that were treated with the negative control should be compared with those of untreated samples.

High transfection efficiencies are required in an RNAi screen because low transfection efficiencies lead to reduced target gene knockdown, reduced phenotypic effects, and lower reproducibility. Consequently, cells must be in optimal physiological condition, and various transfection parameters should be optimized. The commonly considered parameters to be optimized for obtaining maximal transfection efficiency and low toxicity are cell density at the time of transfection, transfection method, amount of transfection reagent, amount of siRNA, and time of treatment/incubation.

In many cases, a pilot study is needed to optimize transfection parameters. For example, before the formal start of a genome-scale RNAi screen, one may conduct a pilot study to optimize several parameters simultaneously so that the maximal transfection efficiency with a low cytotoxicity can be achieved; the parameters might include transfection agents (e.g., from two different vendors), lipid amounts, cell density, siRNA amounts, and incubation time. A negative control such as luciferase may be used to indicate cytotoxicity of delivery methods, and PLK1 might be used as a positive control to measure transfection efficiency.

2.5 Design of Sample Size

In RNAi HTS assays, it is critical to determine a sample size for the achievement of certain false-negative and false-positive levels. The limitation of experimental time and cost usually does not allow a single experiment to have more than two hundred

384-well plates, whereas two hundred 384-well plates is usually the minimal requirement for conducting a genome-wide screen with replicates. Therefore, currently, a typical RNAi HTS project starts with a primary screen of approximately 20,000 siRNAs, most of which have no replicate. The siRNAs identified in the primary screen are further investigated using one or more confirmatory screens in which each siRNA has replicates. A typical primary screen has fifty to one hundred fifty 384-well plates, and a typical confirmatory screen has three to twenty 384-well plates. In the primary screen, we may use the negative control in a plate as a negative reference. The question is: how many wells should we arrange for the negative control in a plate so that we can maintain a manageable false-positive rate while maintaining a reasonably low false-negative rate? Currently, a negative control usually occupies 4, 8, 16, 20, or 24 wells per plate in a primary screen with 384-well plates. Are these numbers enough to maintain low false-positive and false-negative rates? A typical confirmatory screen usually has triplicates for each investigated siRNA. Are triplicates enough to achieve low false-positive and false-negative rates?

To address the questions regarding sample size, we need to conduct a formal statistical analysis to explore the false-positive and false-negative rates across various sample sizes. Statistical analysis for determining sample size in RNAi screens is discussed in Section 5.4 of Chapter 5 and Section 6.5 of Chapter 6. Here I provide guidance for sample size designs that comes from that analysis. In a primary screen using 384-well plates, an arrangement of 4 or 8 wells per plate is not enough to achieve an acceptably low false-negative rate; an arrangement of 16 wells per plate is acceptable, and an arrangement of 20 or 24 wells per plate is preferable for the negative control to be used as a negative reference for hit selection. In a confirmatory screen, a sample size of at least four for each siRNA (i.e., the design of at least four replicate plates per source plate) is required for detecting siRNAs with strong, fairly strong, or moderate effects. Regarding tradeoff between benefit and cost, any sample size between 4 and 11 is a reasonable choice for selecting siRNAs with strong, fairly strong, or moderate effects. If the main focus is the selection of siRNAs with strong effects, a sample size of four or five is a good choice. If cost is not as much of an issue, a sample size of six, seven, or eight is preferred, especially when only one or two sets of source plates are investigated in a confirmatory screen. If enough power is needed to detect siRNAs with moderate effects, then the sample size needs to be 8, 9, 10, or 11 [175].

2.6 Conclusions

As has been illustrated in this chapter and will be demonstrated in the following chapters, experimental design may affect the results of both quality control and hit selection in RNAi screens. The adoption of suitable experimental designs is critical in RNAi screens. With good experimental design, experiments with lower quality or systematic errors can be salvaged. On the other hand, if experimental design is poor, we may not address the questions of interest or salvage experiments that are otherwise good quality. Therefore, the importance of experimental design in a genome-scale RNAi screen cannot be emphasized enough. Consequently, RNAi screening experiments should always be designed appropriately in the beginning to avoid an unpleasant situation that potentially exists in high-throughput biotechnologies, as pointed out by John Quackenbush [43], in which "People tend to go out blindly and do experiments, then go back and try to analyze them and figure out what the question is afterwards."

3

Data Display and Normalization

One of the major advantages of HTS technologies is their ability to simultaneously interrogate thousands of genes/compounds and generate large amounts of data per experiment. To glean biological information from large volumes of data, the first step in data analysis is to use specific graphics to visualize the data and display important features of data. Data display allows the identification of potential problems such as row and column effect, pin issues, and so forth, as they occur [20;58;99;166;173]. If the identified spatial effects are caused by systematic experimental error, we need to adjust for them [20;83;166;173]. Otherwise, they will produce misleading results in both quality control and hit selection. In this chapter, I present graphics for displaying data and explore analytic methods for identifying and/or adjusting for spatial effects that are caused by systematic experimental error. The commonly used analytic methods such as z-score and t-test are based on normal distributions or at least symmetric distributions with constant variance. However, the raw values from RNAi screens are usually skewed with unequal variance. Data transformation is one of the most effective techniques for handling this issue. I also explore data transformation in this chapter.

3.1 Data Display Using Graphics

3.1.1 Plate-Well Series Plot

In a typical RNAi HTS experiment, there are tens to hundreds of plates, each with 384 or 96 wells in which siRNAs are transfected. A scatter plot, called a *plate-well series plot* [180], was designed to display the measured or calculated values well by well and plate by plate in an experiment. In a plate-well series plot, the value of the x-axis is the index of the position of a well in a plate, whereas the labels in the x-axis are the plate number, instead of the index of the position of a well. The positions of wells in a plate can be indexed by either the rows or columns in a plate. In an experiment with n rows and m columns in each plate, if the well is indexed by the rows, the value of x for the well in the jth row and kth column of the ith plate is

$(i − 1)nm + (j − 1)m + k$. Points with x value of 1 to nm denote the nm wells row by row in the first plate; $nm + 1$ to 2 nm denote the nm wells row by row in the second plate, and so on (see Figure 3.1A and B). Similarly, if the well is indexed by the columns, the value of x for the well in the jth row and kth column of ith plate is $(i − 1)nm + (k − 1)n + j$; points with x value of 1 to nm denote the nm wells column by column in the first plate; points with x value of $nm + 1$ to 2 nm denote the nm wells column by column in the second plate; and so on (see Figure 3.1C). The y-axis denotes the intensity either in the original scale, a transformed scale such as log-transformed, or a calculated value such as fold change, percent inhibition, or z-score. When controls are arranged in a plate, we may use different colors or point types to display the values of various control wells in a plate.

The plate-well series plot can also be used to display the results of hit selection in a screen. For example, the sample wells for selected hits are labeled with one color, and the sample wells for non-hits are labeled with another color. See Figure 3.1D for an example showing the use of a plate-well series plot to display hits selected using the criterion of $|SSMD| > 1.645$ (see Chapters 5 and 8 for more details about SSMD).

The advantage of a plate-well series plot is that it can effectively display plate-to-plate variability, show selected hits for all plates, and present common data features of multiple plates in a single plot. For example, the plate-well series plot in Figure 3.1A clearly indicates that the measured values in the first nine plates differ from those in the remaining plates. The measured values in the sample wells and the weaker inhibition control wells in plates 1 to 9 shift down, whereas the measured values in the stronger positive control wells in plates 79 to 88 shift down. The plate-well series plots in Figure 3.1B and C reveal that the measured values in the edge rows are clearly lower than those in the middle rows in each plate of the second confirmatory screen. On the other hand, the well-series plot cannot display the positional effects in an individual plate as straightforwardly as the plate image described in Section 3.1.2.

3.1.2 Plate Image Plots Incorporating Boxplot and Heat Maps

In a genome-scale RNAi screen, each plate may have its own unique data feature; thus it is necessary to check the data plate by plate. The most straightforward approach for displaying data in each individual plate is the use of image of data in the plate [20;166;173]. One common plate image is the so-called *heat map*, in which green represents low values and red represents high values. However, the regular heat map is dominated by extreme data values. In other words, the extremely high or low values such as the outliers make most wells have one color and thus prevent the plate image from accurately displaying data features in the plate (Figure 3.2A). It is well-known that strong true hits behave like outliers and outliers commonly exist in genome-scale RNAi screens. Therefore, the standard heat map or plate image does not work effectively in genome-scale RNAi screens. The regular image plot for a plate in the HCV RNAi second confirmatory screen described in Section 1.4 is shown in

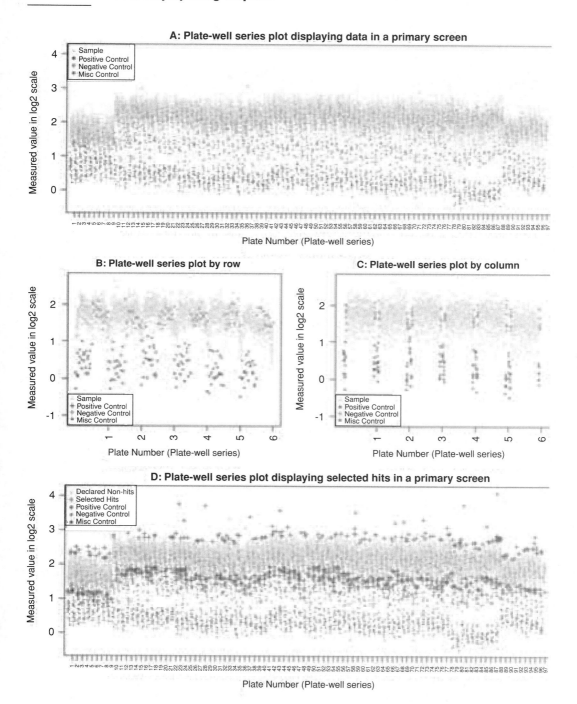

Figure 3.1   (See color insert following page 110.) Plate-well series plots to display data or hits in the HCV RNAi primary screen (A, D) and the second confirmatory screen (B, C) described in Section 1.4. In the legends, Background, Positive Control, and Misc Control denote empty wells, stronger inhibition control wells, and weaker inhibition control wells, respectively.

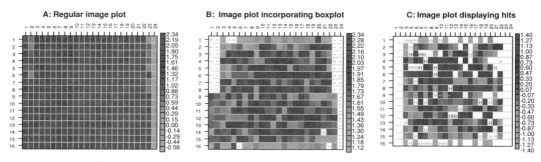

Figure 3.2   (See color insert following page 110.) Regular image plot (Panel A), improved image plot (Panel B) to display the measured value in a plate, and improved image plot to display selected hits in a source plate (Panel C) in the HCV RNAi second confirmatory screen described in Section 1.4 in Chapter 1. The improved image plot clearly reveals a systematic spatial effect (low values in the edge rows and high values in the middle rows), whereas the regular image plot does not reveal the spatial effect.

Figure 3.2A. This image plot barely reveals any data features in the sample wells in that plate because all the sample wells are red.

To address this issue in the regular image plot, we can incorporate *boxplot* statistics into the image plot. That is, we first use the boxplot technique to find the lower and upper whiskers. The whiskers are used to define outliers. Then the strongest green represents the lower whisker (instead of the minimal value in the regular image plot) and the strongest red represents the upper whisker (instead of the maximal value in the regular image plot). A green "–" in a white well indicates that the value in that well is an outlier in the lower end, and a red "+" in a white well indicates that the value in that well is an outlier in the upper end. The values in the legend are those between the upper and lower whiskers. The improved image plot for a plate in the HCV RNAi second confirmatory screen clearly reveals a systematic spatial effect: low values in the edge rows and high values in the middle rows (Figure 3.2B).

Pins are used to transfer liquids to assay plates, and it is not uncommon for RNAi screens to have pin problems. The improved image plot can also help to locate any such pin problems. For example, if some sample wells with the same positions have extremely low (or high) values in all plates or in the majority of the plates in an experiment, then the pins corresponding to these wells potentially have problems. If some well positions have missing values in all plates, then the pins corresponding to these well positions may be missing.

The image plot can also be used to display the positions of selected hits in a plate. To do so, the strongest green color represents the cutoff for selecting hits in the lower end, and the strongest red color represents the cutoff for selecting hits in the upper end. A green "H" in a white well indicates that the siRNA in that well is a selected hit in the lower end, and a red "H" in a white well indicates that the siRNA in that well is a selected hit in the upper end. The values for non-hits are represented by green and red, depending on how large the values are. See Figure 3.2C for an example of

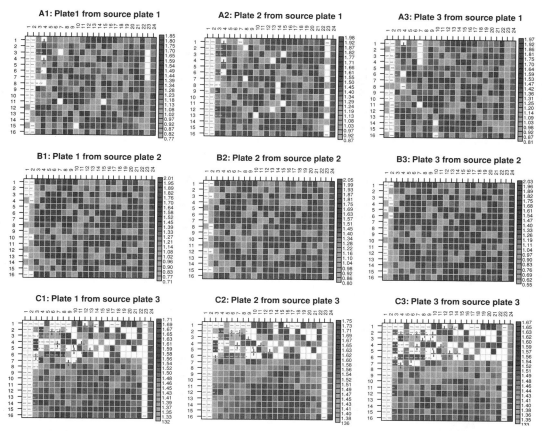

Figure 3.3   (See color insert following page 110.) Improved image plots to display a high reproducibility of data in plates from three source plates (source plates 1, 2, and 3) in an RNAi confirmatory screen. The data in the plates from the same source plate have a very similar pattern, whereas the data in the plates from different source plates have very different patterns.

using image plots to display selected hits using the criterion of |SSMD| > 1.4 (SSMD is discussed in Chapter 4). In this example, the color represents estimated SSMD value. This image plot shows that most up-regulated hits (red "H") are located in the middle, and most down-regulated hits (green "H") are located in the edge of the source plate, which also suggests the existence of edge positional effects in the screen.

Another important use of the improved image plot is to display the reproducibility among replicates of the same source plate; that is, whether the plates from the same source plate have the same pattern of data and the plates from different source plates have a different pattern of data. For example, the improved plate image plot clearly indicates that the reproducibility of data in plates from the same source plate is very high in a confirmatory screen (Figure 3.3).

The so-called *correlation plot* is commonly used to check for reproducibility among replicate plates from the same source plate. A correlation plot is a scatter

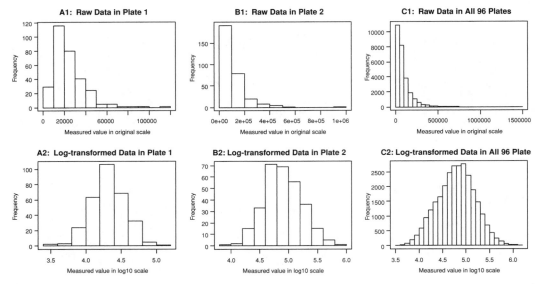

Figure 3.4   Histograms to display the distributions of data in the sample wells in a primary screen with 96 plates. Clearly, the log-transformation makes the skewed data closely symmetric and approximately normal.

plot in which the x-axis represents the values in one plate and the y-axis represents the values in another plate from the same source plate. If there are no systematic spatial effects, correlation plots work effectively. However, if there are unadjusted systematic spatial effects, correlation plots can produce misleading results because the stronger the systematic spatial effects, the higher the correlation between two plates. In other words, a high correlation between two replicate plates may simply indicate a strong spatial effect. Therefore, it is essential to check systematic spatial effects before applying correlation plots for reproducibility.

3.2 Transformation of Measured Raw Values

Most statistical methods for hit selection, including the classical z-score and t-statistic, work best when the data have a normal distribution or at least have a symmetric distribution. Many statistical methods also require the condition of constant variance. That is, the variance for siRNAs with different means of measured values should theoretically have about the same variance. The measured values in their original scale are usually highly skewed to the right (e.g., Figure 3.4A1, B1, and C1). Some measured raw values such as the cell counts theoretically have a Poisson distribution, with variances being linearly linked with means. Therefore, before we formally conduct statistical analysis for quality control and hit selection, we may need to transform the measured raw values.

Detailed analysis about data distributions can be conducted using more specific analytic methods and plots, such as the Shapiro-Wilk test [135], the Anderson-Darling test [2], the Kolmogorov-Smirnov test [140], and QQ plots [10;30]. If a transformation can make the distributions symmetric or approximately normal in most plates, this transformation should work in this screen.

The commonly used transformation is log-transformation. We might also use a square-root transformation for some measured raw values with a Poisson distribution. However, even for cell counts, a log-transformation may work well because the measured raw values are usually large ($\geq$ 5 digits; see Figure 3.4). Therefore, in most cases, we must apply log-transformation to the measured values in their original scale.

Data transformation may have a huge impact on the result of hit selection. Not adopting a suitable transformation may lead to completely misleading results in both quality control and hit selection. For example, in an HIV RNAi primary screen, in which the main objective was to search for down-regulated hits (i.e., inhibition hits), as described in Zhou et al [183], even after we applied the z^*-score method (a robust version of the z-score method) on a plate-by-plate basis, the selected hits based on the raw data were dramatically different from those based on log-transformed data. Figure 3.5 shows that the results based on the log-transformed data are more reasonable than those based on raw data.

3.3 Identification and Adjustment of Systematic Spatial Effects

3.3.1 Identifying Systematic Spatial Effects

Systematic spatial effects commonly exist in RNAi screens. As described in Section 2.3.2, four common patterns of systematic spatial effects are (i) linear row effects, (ii) linear column effects, (iii) linear row and column effects, and (iv) bowl-shaped spatial effects. Spatial effects can be revealed using the improved plate image plot and/or the well-series plot.

An improved image plot can effectively display strong and weak systematic spatial effects in an individual plate (as illustrated in Figure 3.6A1, B1, C1, and D1). It can also reveal various types of spatial effects. Its drawback is that it cannot display quickly the common pattern of spatial effects in many plates. Because each plate requires an image plot, if there are hundreds of plates in an experiment, it may be tough to check them all using the plate-by-plate method. The plate-well series plot can effectively display strong spatial effects in multiple plates (as demonstrated in Figure 3.1B and C). It may also display strong spatial effects in an individual plate, as illustrated in Figure 3.6A2, B2, C2, and D2. However, the plate-well series plot may not reveal weak spatial effects or irregular spatial effects. In addition, for some patterns of systematic spatial effects, such as those in pattern iii (linear row and column effects), we may need one figure plotting against rows and another figure

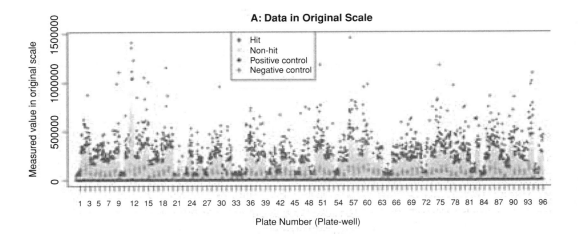

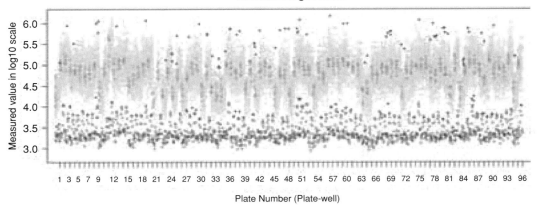

Figure 3.5   (See color insert following page 110.) Plate-well series plots to display the data and selected hits based on original scale (A) or log-transformed scale (B). The use of the criterion of selecting the sample siRNAs with absolute z^*-score (robust version of z-score) greater than 3 as hits leads to the selection of 1,843 up-regulated hits and zero down-regulated hits based on the raw data (A, red points), but selection of 77 up-regulated hits and 29 down-regulated hits based on the log-transformed data (B, red points).

plotting against columns for the same set of plates. It is a good strategy to use both improved image plots and plate-well series plots to check systematic spatial effects.

3.3.2 Consequence of Unadjusted Spatial Effects

Systematic spatial effects can be caused by systematic experimental errors such as liquid evaporation. If systematic spatial effects caused by systematic experimental errors are present, we must adjust for them before conducting analysis for selecting hits. Otherwise, the selected hits will be dominated by spatial effects. That is, whether

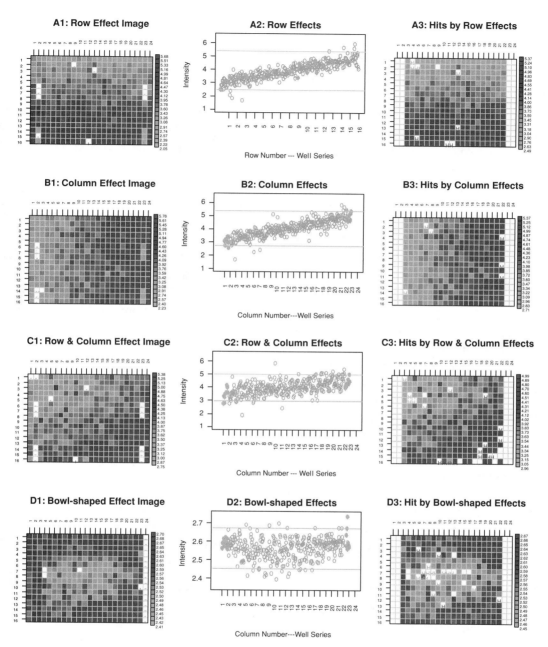

Figure 3.6   (See color insert following page 110.) The use of improved image plot and plate-well series plot to display four common patterns of spatial effects: (i) linear row effects (A1–A3), (ii) linear column effects (B1–B3), (iii) linear row and column effects (C1–C3), and (iv) bowl-shaped spatial effects (D1–D3). In A1, B1, and C1, a red "+" (or a green "–") denotes an outlier in up-regulated (or down-regulated) direction. In A2, B2, and C2, an orange (or green) point denotes a value in a sample well (or a negative control well); the grey lines denote the boundary of SSMD = ±1.4 for selecting hits. In A3, B3, and C3, a red (or green) "H" denotes a selected hit in up-regulated (or down-regulated) direction.

an siRNA is selected as a hit or not is mainly based on its well position in the plate, not based on its knockdown effects.

For example, the selected hits in plates with linear row effects mainly come from the edge columns: down-regulated hits from one edge and up-regulated hits from the other (Figure 3.6A3). Similarly, the selected hits in plates with linear column effects mainly come from the end columns for sample wells (Figure 3.6B3). The selected hits in plates with linear row and column effects mainly come from the two opposite corners (Figure 3.6C3). The selected hits in plates with edge effects, including the bowl-shaped effects, mainly come from the middle wells and edge wells (Figure 3.6D3). These results distort the true pattern.

Spatial effects caused by systematic experimental errors may also increase the variability of the data we use to set up the boundary for selecting hits. Linear row and/or column effects may substantially increase the value of the median of absolute deviations (MAD) when we use the median $\pm 2 \times$ MAD method for selecting hits (median $\pm 2 \times$ MAD methods are described in detail in Section 5.2). As a result, the upper boundary of median $+ 2 \times$ MAD is too high, and the low boundary of median $- 2 \times$ MAD is too low; subsequently, we will miss more true hits and have a higher false-negative rate. In Figure 3.6, for example, without adjusting for spatial effects, only a few hits are identified in the first three plates (Figure 3.6A3, B3, and C3). However, after adjustment for spatial effects, many more hits are identified (Figure 3.7A3, B3, and C3).

Systematic spatial effects can also produce misleading quality control results in both primary and confirmatory screens [166;173]. See Chapter 4 for more details about the impact of systematic experimental errors on quality control.

3.3.3 Methods for Adjusting for Systematic Spatial Effects

There are several statistical methods for addressing systematic spatial effects. The first is regular linear regression, in which we fit a linear model of the measured value (after suitable transformation) of an siRNA in a well over the mean of values in its row, the mean of values in its column, and, possibly, over the interaction between the row and column. A simple formula for the linear model is:

$y \sim$ row.mean $+$ column.mean $+$ row.mean $*$ column.mean

The adjusted value for an siRNA is its measured value minus its fitted value plus the mean of all fitted values.

To address the common problem of outliers in HTS experiments, we may use robust linear models. One such model is median polishing, in which we fit a linear model of the measured value (after suitable transformation) of an siRNA in a well over the median of values in its row, the median of values in its column, and, possibly, over the interaction between the row and column. A simple formula for median polishing is:

$y \sim$ row.median $+$ column.median $+$ row.median $*$ column.median

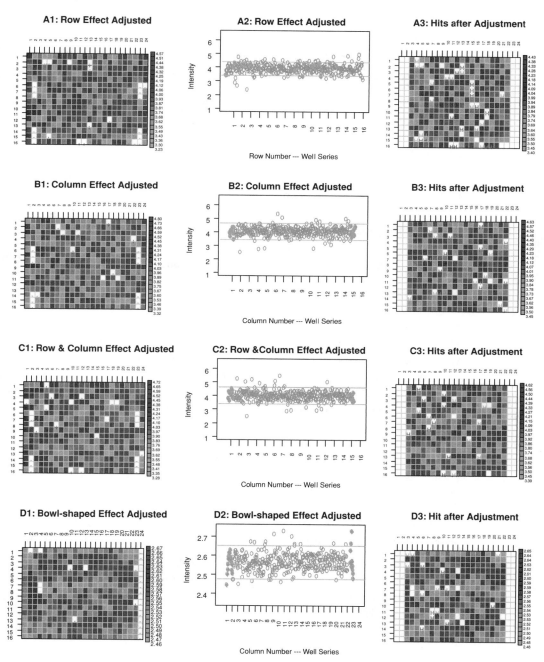

Figure 3.7 (See color insert following page 110.) Adjustment of four common patterns of spatial effects: (i) linear row effects (A1–A3), (ii) linear column effects (B1–B3), (iii) linear row and column effects (C1–C3), and (iv) bowl-shaped spatial effects (D1–D3). In A1, B1, and C1, a red "+" (or a green "–") denotes an outlier in up-regulated (or down-regulated) direction. In A2, B2, and C2, an orange (or green) point denotes a value in a sample well (or a negative control well); the grey lines denote the boundary of SSMD = ±1.4 for selecting hits. In A3, B3, and C3, a red (or green) "H" denotes a selected hit in up-regulated (or down-regulated) direction.

In a regular linear model, we usually find the fitted value by minimizing mean squared error, whereas in median polishing, we find the fitted value by minimizing the median absolute deviation. The median polishing method can be implemented using B-scores [20]. A more convenient choice may be the use of robust regression implemented in the R function *rlm* [71;154].

The basic R codes using function *rlm* to adjust positional effects are as follows. For adjusting both row and column effects:

```
library(MASS)
Yraw = dataIn.df[, "Intensity"]
Yfits = rlm(Intensity ~ Xpos * Ypos, data = dataInNeg.df)
# or simply Yfits = rlm(Intensity ~ Xpos + Ypos, data = dataInNeg.df)
Ypredicted = predict(Yfits, data.frame("Xpos"=dataIn.df[, "Xpos"],
  "Ypos"=dataIn.df[, "Ypos"]))
Yadjusted = Yraw - Ypredicted + mean(Ypredicted)
```

For adjusting row effects only:

```
Yraw = dataIn.df[, "Intensity"]
Yfits = rlm(Intensity ~ Xpos, data = dataInNeg.df)
Ypredicted = predict(Yfits, data.frame("Xpos"=dataIn.df[, "Xpos"]))
Yadjusted = Yraw - Ypredicted + mean(Ypredicted)
```

For adjusting column effects only:

```
Yraw = dataIn.df[, "Intensity"]
Yfits = rlm(Intensity ~ Ypos, data = dataInNeg.df)
Ypredicted = predict(Yfits, data.frame("Ypos"=dataIn.df[, "Ypos"]))
Yadjusted = Yraw - Ypredicted + mean(Ypredicted)
```

In the above codes, *dataIn.df* is a data frame containing the data for all wells in a plate, which must contain three columns: *Xpos*, *Ypos*, and *Intensity*. These three columns represent, respectively, the row number, column number, and measured response value (usually in log scale) for each well. *dataInNeg.df* is a part of *dataIn.df* for data in the negative reference wells only. The negative reference can be either all negative control wells if the negative control is arranged across the plate or all sample wells (see Section 5.2.1 of Chapter 5 for a more detailed discussion about the negative reference in hit selection).

The linear model and its robust version work effectively when the spatial effect is linear. However, to address for nonlinear spatial effects, such as bowl-shaped effects, we need to use smoothing techniques such as local fitting [31], smoothing splines [56], and some regular nonlinear models to find a smoothing surface over the well positions. The adjusted value for an siRNA in a well is the measured value minus its corresponding value in the smoothing surface in that well plus the mean of all fitted values in the smoothing surface. It is not easy to determine which regular nonlinear model to apply; thus local fitting and smoothing splines are commonly used. A convenient and usually effective method is to use the function *loess* or sometimes *smooth.spline* in R to find the fitted value [166;173].

The basic R codes using function *loess* to adjust positional effects are:

```
Yraw = dataIn.df[, "Intensity"]
Yfits = loess(Intensity ~ Xpos * Ypos, dataInNeg.df,
   control = loess.control(surface = "direct"))
Ypredicted = predict(Yfits, data.frame("Xpos"=dataIn.df[, "Xpos"],
   "Ypos"=dataIn.df[, "Ypos"]))
Yadjusted = Yraw - Ypredicted + mean(Ypredicted)
```

3.3.4 Applications in Adjustment of Systematic Spatial Effects

After choosing an analytic method for adjusting for spatial effects that are caused by systematic experimental errors, we need to determine which type of wells to use for fitting the model. In a primary screen in which the siRNAs are randomly arranged in a plate, the majority of siRNAs in a plate should not have a large inhibition or activation effect. If they do, we may use the sample wells to fit the model and then use the fitted model to find fitted values for every well, including the control wells in a plate.

For example, the four plates displayed in Figure 3.6 come from primary screens, although each one is from a different screen. Thus we can use the 320 sample wells to fit the model. For each of the first three plates, robust regression can be used to build the model based on the rows and columns of all the 320 sample wells; the fitted values for all 384 wells can then be found in a plate. The adjusted values for these three plates using R function rlm are displayed in Figure 3.7A1 and A2, B1 and B2, and C1 and C2, respectively. The fourth plate can be adjusted using smoothing based on all 320 sample wells. The adjusted values for this plate using R function loess are displayed in Figure 3.7D1 and D2. Figure 3.7 clearly indicates that the spatial effects in all the four plates have been adjusted. Figure 3.7A3, B3, C3, and D3 indicate that the impact of systematic experimental errors on hit selection has been removed.

In a confirmatory screen in which most siRNAs are selected from primary screens and are supposed to have large or moderate effects, sample wells cannot be used to fit the model because the systematic spatial effects may be caused by clusters of true hits in an area of a plate. Fitting a model to adjust for spatial effects may substantially reduce the measured signal of true hits. For example, we do not want to use local fitting on sample wells to remove the spatial pattern in Figure 3.3B1, B2, and B3 or C1, C2, and C3 because those spatial patterns are caused by the location of true hits in those plates.

Thus the best strategy to adjust for spatial effects in a confirmatory screen is to use well-designed negative control wells or a control that is supposed to

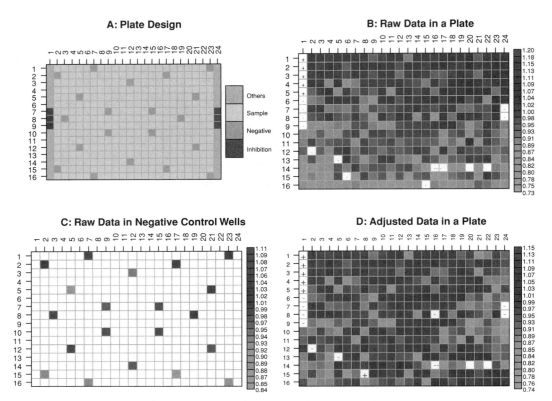

Figure 3.8   (See color insert following page 110.) Adjustment of spatial effects using an effective plate design in a confirmatory screen.

consistently have the same or a similarly sized effect in an experiment. Unfortunately, the arrangement of negative control wells in a plate is not well designed in many experiments. Therefore, following the strategy of plate designs described in Section 2.3 of Chapter 2 is critical for the adjustment of systematic spatial effects caused by systematic experimental error. For example, both the plate-well series plot (Figure 3.1B) and the image plot (Figure 3.2B) show that there are convex bowl-shaped spatial effects in the second HCV screen. However, because all the negative control wells were arranged in column 23, it is infeasible to determine whether these spatial effects are caused by systematic error or the arrangement of sample wells.

An example of adopting effective plate design to adjust for spatial effects is an siRNA confirmatory screen in which the negative control wells are arranged as in Figure 3.8A. The display of measured values indicates the existence of systematic row effects (Figure 3.8B). The values in the negative control wells indicate that the systematic spatial effects were caused by measurement error (Figure 3.8C). Based on the values in the negative control wells shown in Figure 3.8C, we can adjust for the spatial effect in Figure 3.8B to get the adjusted data shown in Figure 3.8D.

3.4 Strategy for Data Display and Normalization

Data display and normalization represent the first step for data analysis in RNAi screens. Data display may identify important data features, including systematic experimental errors in an RNAi screening experiment. Transformation of measured raw values and adjustment of systematic experimental errors may have a huge impact on the results of hit selection (Figures 3.5 through 3.8). A good strategy for data display and normalization is to follow the following steps: (i) determine what transformation should be applied to the measured raw values in a screen; (ii) use graphics to display data features and to examine whether any systematic spatial effects exist; (iii) if systematic spatial effects exist, explore whether they are caused by systematic experiment errors, such as evaporation, pin issues, and so forth; (iv) if the existing systematic spatial effects are caused by systematic experimental errors, adopt a suitable analytic method such as robust linear model and local fitting to adjust for them.

The best strategy to adjust for spatial effects in a confirmatory screen is to use well-designed negative control wells or a control that has consistently the same or similar size effects. If all controls are not well designed for the adjustment of spatial effects in a confirmatory screen, we may adopt the following strategy: (i) if many plates in a screen have the same pattern of spatial effects, regardless of whether they come from the same or different source plates, we may apply the corresponding adjustment method based on the sample wells in each plate that has systematic spatial effects; (ii) if only plates from one or two source plates have certain patterns of systematic spatial effects, we may not adjust for the spatial effects because the spatial effects may be caused by the positions of true hits in those source plates.

4

Quality Control in Genome-Scale RNAi Screens

4.1 Introduction

High-quality RNAi HTS assays are critical in genome-scale RNAi research. The development of high-quality RNAi HTS assays requires the integration of both experimental and computational approaches for quality control (QC). Three important means of QC are (i) good plate design, (ii) the selection of effective positive and negative biological controls, and (iii) the development of effective QC metrics to measure the degree of differentiation so that assays with inferior data quality can be identified.

Plate design and the design of effective controls are described in Chapter 2. A good plate design helps to identify systematic errors (especially those linked with well position) and determine what normalization should be used to remove/reduce the impact of systematic errors on both QC and hit selection. Section 2.3 presents multiple effective plate designs and guidelines; more information is available in Zhang [166]. In this chapter, the development of effective QC metrics and the use of effective QC criteria are discussed, and the use all three QC processes to improve data quality in genome-scale RNAi screens is demonstrated.

4.2 Quality Assessment Metrics

Effective analytic QC methods serve as a gatekeeper for excellent quality assays. In a typical HTS experiment, a clear distinction between a positive control and a negative reference such as a negative control is an index for good quality. Many quality assessment measures have been proposed to measure the degree of differentiation between a positive control and a negative reference. Signal-to-background ratio, signal-to-noise ratio, signal window, assay variability ratio, and Z-factor have been adopted to evaluate data quality [17;39;77;99;116;123;148;150;159;172;180;185]. *Strictly standardized mean difference* (SSMD) has recently been proposed for assessing data quality in RNAi HTS assays [162]. The commonalities and differences of these measures are shown in the formulas for their estimates in Table 4.1.

Table 4.1. Quality control measures, their estimation formulas, and characteristics

QC Measure	Estimation Formula	Characteristics
Signal-to-background ratio	$\dfrac{\bar{X}_P}{\bar{X}_N}$	Interpretation based on graphics Does not contain any information regarding data variability
Signal-to-noise ratio	$\dfrac{\bar{X}_P - \bar{X}_N}{s_N}$	Interpretation based on graphics Takes into account the variability in the negative control but not in the positive control
Signal window	$\dfrac{\lvert \bar{X}_P - \bar{X}_N \rvert - 3(s_P + s_N)}{s_N}$	Accounts for data variability in both negative and positive controls Interpretation based on graphics
Assay variability ratio	$\dfrac{3(s_P + s_N)}{\lvert \bar{X}_P - \bar{X}_N \rvert}$	Accounts for data variability in both controls Interpretation based on graphics Assay variability ratio = 1 + Z-factor
Z-factor	$\dfrac{\lvert \bar{X}_P - \bar{X}_N \rvert - 3(s_P + s_N)}{\lvert \bar{X}_P - \bar{X}_N \rvert}$	Accounts for data variability in both controls Interpretation based on graphics No direct probability-based interpretation Difficult to derive its estimation and confidence interval from a complete statistical basis
SSMD	$\dfrac{\bar{X}_P - \bar{X}_N}{\sqrt{s_P^2 + s_N^2}}$	Accounts for data variability in both controls Interpretation based on probability: has a direct relationship with the probability that a draw from the positive control is greater than a draw from the negative control Simple to derive its estimation and confidence interval from a complete statistical basis

Note: QC, quality control; $\bar{X}_P$ and s_P are sample mean and standard deviation of a positive control, respectively, and $\bar{X}_N$ and s_N are sample mean and standard deviation of a negative reference, respectively.

All these methods have advantages and drawbacks. From Table 4.1, we see that signal-to-background ratio is simple to calculate but does not contain any information regarding data variability, because it does not include the standard deviation. Signal-to-noise ratio takes into account variability in the negative reference but not that of the positive control. Consequently, QC results reached using signal-to-background ratio and signal-to-noise ratio are misleading [17;77;159;162]. For example, the data distributions and observed values in Figure 4.1 clearly show that the degrees of differentiation between the positive and negative controls are very different from cases A to D. However, the signal-to-background ratios are the same for all four cases, and the signal-to-noise ratios are the same for cases A through C.

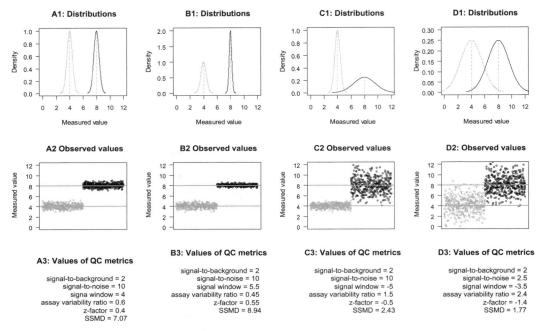

Figure 4.1   Distributions and observed values of positive controls (marked as black curves and points) and negative controls (marked as grey curves and points) and the corresponding values of QC metrics in four cases: Case A (A1–A3), B (B1–B3), C (C1–C3), and D (D1–D3). The grey straight lines denote the means in the positive control and the negative reference, respectively.

Signal window, assay variability ratio, Z-factor, and SSMD all capture data variability. For a signal window, a larger value should indicate a larger degree of differentiation between the positive control and the negative control. The value of signal window is larger in case D than in case C. Thus the use of signal window for QC would have led to the conclusion that the quality of case D is better than that of case C. However, the distributions in C1 and D1 reveal that the overlap between distributions in the two controls in case C is smaller than in case D. The observed values in C2 and D2 also indicate that more values in the positive controls in case C separate from those in the negative control than in case D. Hence the distributions and observed values clearly reveal that the degree of differentiation in case C is larger than in case D; consequently, the quality of case C is better than that of case D (Figure 4.1). Therefore, signal window also leads to misleading results. The table clearly shows that the assay variability ratio $= 1 + Z$-factor; thus the assay variability ratio and Z-factor are equivalent. Using simulations, Iversen et al. [77] show that Z-factor is better than signal window in terms of accuracy and precision.

I recommend using SSMD whenever possible. SSMD and signal-to-noise ratio have the same numerator but different denominators. Signal-to-noise ratio incorporates variability only in the negative reference, whereas SSMD accounts for variability in both controls. Consequently, when the two controls have different variability,

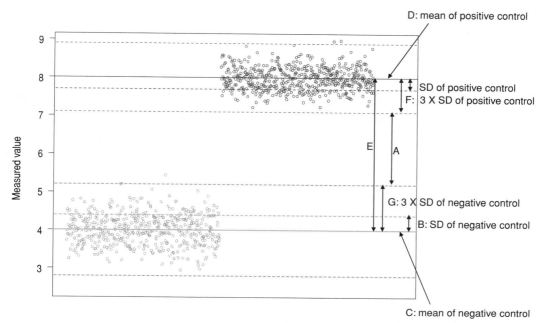

Figure 4.2   Graphic interpretation of quality control metrics. Signal-to-background ratio = D/C; signal-to-noise ratio = E/B; signal window = A/B; assay variability ratio = (F + G)/E; Z-factor = A/E. The black and grey points denote observed values in a positive control and a negative reference, respectively.

signal-to-noise ratio may give misleading results, whereas SSMD does not. SSMD is similar to the reverse of the assay variability ratio. Both take into account variability in both controls; however, the methods differ in terms of how they combine information about this variability: assay variability ratio and Z-factor directly sum up the standard deviations in the two controls, whereas SSMD adopts the standard deviation of the difference between the positive and negative controls. As a result, assay variability ratio and Z-factor are conveniently interpreted using graphics, or control chart (as shown in Figure 4.2), based on means and standard deviations in a similar way to signal-to-background ratio, signal-to-noise ratio, and signal window, whereas SSMD is better in terms of both probabilistic interpretation and statistical estimation and inference, as elaborated in the following two paragraphs. This is an advantage of SSMD because of the stochastic nature of readouts in HTS assays.

Probabilistic basis of SSMD. SSMD has a direct relationship with d^+-probability; that is, the probability that a draw from the positive control is greater than a draw from the negative reference (see Section 8.1 of Chapter 8). The simplest relationship is d^+-probability = Φ(SSMD), where $\Phi(\cdot)$ is a cumulative distribution function of a standard normal distribution $N(0, 1)$. This relationship offers a strong basis for a probabilistic interpretation of SSMD-based criteria. By contrast, there is no direct relationship between Z-factor and related probability, although Sui and Wu

[148] roughly give a power interpretation to Z-factor–based criteria under certain assumptions and limitations. SSMD is also easily interpreted by average fold change (in log scale) standardized by the variability of fold change (in log scale) between the two controls.

The fact that the Z-factor is based on an absolute value rather than a normal value also makes it more difficult to derive estimation and inference for Z-factor. By contrast, it is simple to derive the estimation and confidence interval of SSMD from a complete statistical basis. When the positive control and the negative reference have equal variance, an estimate (namely, minimal variance unbiased estimate) of SSMD better than the estimate in Table 4.1 is as follows:

$$
\begin{aligned}
\text{SSMD} &= \frac{\bar{X}_P - \bar{X}_N}{\sqrt{\dfrac{2}{K}\left((n_P - 1)s_P^2 + (n_N - 1)s_N^2\right)}} \\
&\approx \frac{\bar{X}_P - \bar{X}_N}{\sqrt{\dfrac{2}{n_P + n_N - 3.5}\left((n_P - 1)s_P^2 + (n_N - 1)s_N^2\right)}}
\end{aligned}
\tag{4.1}
$$

where n_P and n_N are sample sizes of the positive control and the negative reference, respectively, $n_P, n_N \geq 2$ and $K = 2 \cdot \left(\left(\Gamma\left(\frac{n_P + n_N - 2}{2}\right)\right) \Big/ \left(\Gamma\left(\frac{n_P + n_N - 3}{2}\right)\right)\right)^2 \approx n_P + n_N - 3.5$ (see Chapter 8 details regarding statistical derivation).

t-statistic and p-value. The SSMD estimate in Table 4.1 looks similar to the t-statistic in situations of unequal variance:

$$
t\text{-statistic} = \frac{\bar{X}_P - \bar{X}_N}{\sqrt{\dfrac{s_P^2}{n_P} + \dfrac{s_N^2}{n_N}}}.
\tag{4.2}
$$

The SSMD estimate in Formula 4.1 looks similar to the t-statistic in situations of equal variance:

$$
t\text{-statistic} = \frac{\bar{X}_P - \bar{X}_N}{\sqrt{\dfrac{1}{n_P + n_N - 2}\left((n_P - 1)s_P^2 + (n_N - 1)s_N^2\right)} \cdot \sqrt{\dfrac{1}{n_P} + \dfrac{1}{n_N}}}.
\tag{4.3}
$$

However, there is a major difference between the t-statistic and SSMD. In situations of either equal or unequal variance, if there is a tiny true mean difference, when n_P and/or n_N increases, t-statistics increase and corresponding p-values decrease, whereas SSMD estimates tend to be closer to its population value [162]. In other words, a t-statistic is a function of both sample size of controls and degree of differentiation between two controls and is thus highly affected by sample size of controls, whereas the values of an SSMD estimate fall around SSMD population value. This difference makes the t-statistic a poor metric and makes SSMD a good metric for assessing data quality.

For example, no matter how poor an assay is, one can increase the t-statistic to reach a very large value (or reduce the corresponding p-value to reach a very small value) simply by increasing the sample size of controls in a plate, as long as there is a tiny mean difference between the positive and negative controls. By contrast, increasing the sample size only increases the SSMD estimate if its population value is higher. The population value of SSMD reflects only the degree of differentiation between the two controls and is not affected by the sample size of the controls. The Z-factor should have a similar property to SSMD in terms of sample size impact, which is a major reason why the Z-factor has been much more widely used to assess quality in HTS assays than the t-statistic and its associated p-value.

Statistical power. Statistical power for testing null hypotheses has also been proposed for a QC metric [148]. However, power is a more complicated term than the t-statistic and its corresponding p-value. Power is further affected by type I error as well as the effect size and the sample size of the controls. Thus t-statistic, p-value, and power are all suitable for null hypothesis testing but are not suitable for measuring effect size, especially with the consideration that many siRNAs may have tiny effects on measured response due to gene network (Harlow, Mulaik, and Steiger [65] provide more serious criticisms).

4.3 Quality Control Criteria

SSMD cutoffs. To assess data quality, we should first choose an effective QC metric and then determine cutoffs of the chosen metric for classifying quality type. Considering the advantages of SSMD as described in Section 4.2, we choose SSMD as the QC metric and use the SSMD-based criteria listed in Table 4.2 for QC in RNAi screens [166].

The thresholds $\hat{\beta}$ in the SSMD-based criteria in Table 4.2 have a theoretical basis and a probabilistic interpretation. The SSMD values of 0.5, 1, 2, and 3 have clear meanings: the size of mean difference being one half, one time, two times, and three times the standard deviation of the difference, respectively. The d^+-probability associated with SSMD of an siRNA (or a positive control) is the probability that a value from this siRNA (or the positive control) is greater than a value from a negative reference. Based on the relationship between SSMD and d^+-probability, the SSMD values of 1, 2, and 3 (or -1, -2, and -3) also indicate that the minimums (or maximums) of the corresponding d^+-probabilities are respectively approximately 0.5, 0.95, and 0.975 (or 0.5, 0.05, and 0.025) in a situation in which the difference has a symmetric unimodal distribution with finite variance. Similarly, an SSMD value of 4.7 (or -4.7) indicates that the minimum (or maximum) of the corresponding d^+-probability is 0.99 (or 0.01). In a situation in which the difference has a normal distribution, SSMD $= 0.5, 1, 2, 3, 4.7$ corresponds to d^+-probability $= 0.69, 0.84, 0.97725, 0.99865$, and 0.9999987, respectively. The SSMD-based criteria consider various distributions and are thus robust to different symmetric distributions.

Table 4.2. SSMD-based QC criteria in RNAi HTS assays taking into account effect size of a positive control in an assay and the strategies for choosing an SSMD-based criterion

| | SSMD-Based QC Criteria | | | | | | | |
| | For a Moderate Control | | For a Strong Control | | For a Very Strong Control | | For an Extremely Strong Control | |
Quality	Ia	IIa	Ib	IIb	Ic	IIc	Id	IId
Types								
Excellent	$\hat{\beta} \geq 2$	$\hat{\beta} \leq -2$	$\hat{\beta} \geq 3$	$\hat{\beta} \leq -3$	$\hat{\beta} \geq 5$	$\hat{\beta} \leq -5$	$\hat{\beta} \geq 7$	$\hat{\beta} \leq -7$
Good	$2 > \hat{\beta} \geq 1$	$-2 < \hat{\beta} \leq -1$	$3 > \hat{\beta} \geq 2$	$-3 < \hat{\beta} \leq -2$	$5 > \hat{\beta} \geq 3$	$-5 < \hat{\beta} \leq -3$	$7 > \hat{\beta} \geq 5$	$-7 < \hat{\beta} \leq -5$
Inferior	$1 > \hat{\beta} \geq 0.5$	$-1 < \hat{\beta} \leq -0.5$	$2 > \hat{\beta} \geq 1$	$-2 < \hat{\beta} \leq -1$	$3 > \hat{\beta} \geq 2$	$-3 < \hat{\beta} \leq -2$	$5 > \hat{\beta} \geq 3$	$-5 < \hat{\beta} \leq -3$
Poor	$\hat{\beta} < 0.5$	$\hat{\beta} > -0.5$	$\hat{\beta} < 1$	$\hat{\beta} > -1$	$\hat{\beta} < 2$	$\hat{\beta} > -2$	$\hat{\beta} < 3$	$\hat{\beta} > -3$

Notes: ($\hat{\beta}$ denotes estimated SSMD value.)

1. Criteria Ia, Ib, Ic, and Id are applied to situations in which the intensity of a positive control is theoretically larger than that of a negative reference, and criteria IIa, IIb, IIc, and IId applied to situations in which the intensity of a positive control is theoretically smaller than that of a negative reference. In practice, we usually know whether the measured intensity of a positive control is theoretically larger than a negative reference.

2. If the effect size of a positive control is known biologically, adopt the corresponding criterion based on this table; otherwise, use the following strategies for choosing which SSMD-based criterion to use in an RNAi HTS experiment:

   a) For RNAi HTS assays in which cell viability is the measured response, criterion Id or IId should be adopted for the controls without cells (namely, the wells with no cells added) or background controls.

   b) If the difference is not normally distributed, especially when it is highly skewed, QC criteria Id and IId may be used, even for a strong or very strong positive control.

   c) In a viral assay in which the amount of viruses in host cells is the interest, criterion Ic or IIc is usually used, and criterion Id or IId is occasionally used for the positive control consisting of siRNA from the virus.

   d) If there is not enough information about the positive controls that contain siRNAs:

      i) Adopt criterion Ic or IIc when there is only one positive control in an experiment.

      ii) Adopt criterion Ic or IIc for the stronger positive control and criterion Ib or IIb for the weaker positive control when there are two positive controls in an experiment.

   e) In compound HTS assays, positive controls usually have very strong effects; thus usually criterion Id or IId (and occasionally criterion Ic or IIc) should be adopted in compound HTS assays.

Table 4.3. *Z*-factor-based criteria for classifying assay quality

Category	Z-Factor-Based Criterion
Ideal	Z-factor $= 1$
Excellent	$1 > Z$-factor ≥ 0.5
Doable	$0.5 > Z$-factor > 0
Yes/no	Z-factor $= 0$
Screening essentially impossible	Z-factor < 0

Source: From Zhang, Chung, and Oldenburg [159].

Z-factor criteria. In the early stages of HTS technology research, *Z*-factor, along with the *Z*-factor-based criteria listed in Table 4.3, were widely used for evaluating data quality [17;99;116;123;150;159;160;180;185]. The *Z*-factor-based criteria were derived empirically [159] and lack a clear probabilistic interpretation [161;162;164]. Consequently, we do not know in what conditions they work best and in what situations they have limitations. The clear probabilistic interpretation of SSMD may help to find an indirect probabilistic interpretation of the *Z*-factor-based criteria, as follows.

Relationship between SSMD and Z-factor. Because $\sqrt{s_P^2 + s_N^2} < s_P + s_N$ when $s_P > 0$ and $s_N > 0$, we have $|\text{SSMD}| = |\bar{X}_P - \bar{X}_N| \big/ \sqrt{s_P^2 + s_N^2} > |\bar{X}_P - \bar{X}_N| \big/ (s_P + s_N) = 3/(1 - Z\text{-factor})$. Therefore, given a value z (not greater than 1), if Z-factor $> z$, then $|\text{SSMD}| > 3/(1 - z)$; however, if $|\text{SSMD}| > 3/(1 - z)$, we cannot ensure that Z-factor $> z$. Therefore, Z-factor $> z$ is a subset of $|\text{SSMD}| > 3/(1 - z)$, which leads to the result that Z-factor > 0 is a subset of $|\text{SSMD}| > 3$ and Z-factor > 0.5 is a subset of $|\text{SSMD}| > 6$ [162].

Consequently, the criterion of Z-factor > 0 is more conservative than the criterion of $|\text{SSMD}| > 3$. When the positive control has a mean greater than the negative control, the use of SSMD > 3 already requires that, if we draw two values from the positive and negative controls, respectively, the probability that the draw from the positive control must be greater than the draw from the negative control is greater than 0.99865 under normal assumption. That is, SSMD > 3 is already a strong requirement. The more conservative criterion of Z-factor > 0 (namely, not screening essentially impossible) has an even stronger requirement than SSMD > 3. Thus the popularly used *Z*-factor-based QC criterion is most suitable for very or extremely strong positive controls.

Level of controls. In an RNAi HTS assay, a strong or moderate positive control is usually more instructive than a very or extremely strong positive control because the effectiveness of this control is more similar to HTS hits of interest (see the HCV example in Section 2.2.2 of Chapter 2). Applying the *Z*-factor–based QC criteria in Table 4.3 to strong or moderate positive control may not help to identify true hits of interest. For example, in the simulation study shown in Figure 4.3, the assay quality in each simulated experiment is ideal: data variability consists of biological variability but no assay variability. The positive control in experiment A has stronger inhibition effects than the positive control in experiment B, as shown in A1 and B1. If we use the Z-factor-based QC criteria in both experiments, 28 plates are "screening essentially impossible" and 72 are "doable" in experiment A (A3), and 95 plates are "screening essentially impossible" and 5 are "doable" in experiment B (B3). Thus many plates with good quality are mistakenly judged to have poor quality when using the Z-factor-based criteria.

Because the positive controls in the two experiments theoretically have different sizes of inhibition effects, the QC thresholds for the moderate control should be different from those for the strong control in these two experiments. In addition, it is common that two or more positive controls are adopted in a single experiment, such as in experiment C (Figure 4.3). Applying the same Z-factor-based QC criteria to both controls leads to inconsistent results: 32 "excellent" and 68 "doable" by one positive control, represented by purple points, and 4 "doable" and 96 "screening essentially impossible" by the other control, represented by red points (Figure 4.3C3). Compared with the Z-factor-based criteria, the SSMD-based QC criteria in Table 4.2 have the merit of taking into account the size and direction of effects of positive controls in an HTS assay, thus working effectively to address the need for QC in RNAi screens.

Based on the SSMD-based QC criteria and related strategies, criteria IIc and IIb are applied to the very strong control in experiment A and the strong control in experiment B, respectively, which produces the following sensible QC evaluation results: 21 excellent and 79 good plates in experiment A and 31 excellent, 66 good, and 3 inferior plates in experiment B. Criteria IId and IIb are applied to the extremely strong control and the strong positive control in experiment C, respectively, which yields the following evaluation results: 60 excellent and 40 good plates by the extremely strong control and 34 excellent, 63 good, and 3 inferior plates by the strong positive control. All the QC results obtained using SSMD-based QC criteria are much more reasonable than those obtained using the Z-factor-based QC criteria.

To apply the SSMD-based QC criteria listed in Table 4.2, we need to know the size and direction of effects of positive controls in an HTS assay. In practice, we usually know whether the measured intensity of a positive control is theoretically greater than a negative reference. It is more difficult to obtain the information about the sizes of positive controls. Most positive controls adopted in RNAi HTS experiments have strong or very strong effects. Therefore, the QC criteria Ib, Ic, IIb, and IIc listed in Table 4.2 work for most RNAi HTS experiments. In some cases, a good positive biological control is unavailable for experimenters, and some very strong positive controls must be used. For example, in some experiments in which the goal is to screen out siRNAs capable of inhibiting cancer cell growth, the wells with no cells are used as a positive control because no better positive biological controls are available. Because these positive control wells have no cells added at all, the size

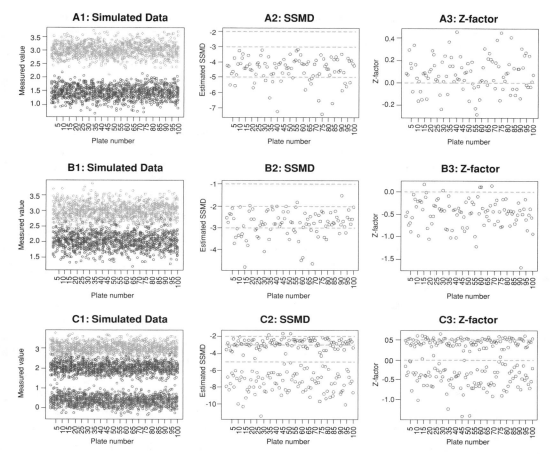

Figure 4.3   (See color insert following page 110.) Data, SSMD, and Z-factor in three simulated experiments A, B, and C in which the positive controls have different effect sizes: a very strong control (red points) in experiment A, a strong control (red points) in experiment B, and an extremely strong control (purple points) and a strong control (red points) in experiment C. In each simulated experiment, there are 100 plates each, with 10 replicates for each positive or negative control. The data for each control in each experiment is generated from a normal distribution with standard deviation of 0.25, namely $N(\mu, 0.25^2)$ where $\mu = 3$ for the negative control in each experiment, and $\mu = 1.44, 2.01, 0.35, 2.01$ for the very strong positive control in experiment A, strong positive control in experiment B, and extremely strong positive control and strong positive control in experiment C, respectively.

of inhibition effect in this positive control is extremely strong. In such a case, we may use QC criterion Id or IId. Detailed strategies for adopting an SSMD-based QC criterion for a positive control in an HTS experiment in which there is not enough information about the size of positive controls are described in the Notes section of Table 4.2.

To use SSMD-based QC criteria for judging whether a plate passes or fails QC in a screen with only one positive control, the following strategy may be adopted: a plate

Table 4.4. Strategies for quality controls in a screen with one or two positive controls

Strategy 1. Using one positive control:
- A plate passes QC if it has good or excellent quality for the positive control.
- A plate fails QC if it has poor or inferior quality for the positive control.

Strategy 2. Using two positive controls:
- A plate passes QC if it has good or excellent quality for both positive controls.
- A plate fails QC if it has inferior or poor quality for both positive controls.
- Depending on experimental need and cost, a plate may pass or fail QC if it has good or excellent quality for one positive control and inferior or poor quality for the other positive control.

passes QC if it has excellent or good quality and fails QC if it has inferior or poor quality. When multiple positive controls are used in an HTS screen, we may need to evaluate data quality based on two or more positive controls. In many cases, to pass QC in a plate, we may need both positive controls to pass QC in that plate, especially in the experiments with the objective of selecting both the siRNAs with very strong effects and those with strong or moderate effects. The strategies for the adoption of SSMD-based criteria for QC in an HTS experiment are summarized in Table 4.4.

Although I recommend the use of different criteria for controls with different effect sizes in RNAi HTS experiments, especially in situations in which there is no replicate for the majority of sample siRNAs in a plate, there are disputes about whether a single criterion or multiple criteria should be used. The adoption of multiple criteria for controls with different effect sizes (such as the SSMD-based QC criteria in Table 4.2) takes into account the fact that different positive controls may have different effect sizes; however, it is complicated to apply them in experiments in which the sizes of positive controls are unknown. A single criterion is simple to apply in experiments; however, it cannot account for the fact that different positive controls may have different effect sizes and thus leads to inconsistent QC results in experiments with two or more positive controls with different effect sizes.

In RNAi HTS experiments with no replicate for sample siRNAs, it is important to allow for the sizes of positive controls because (i) a moderate or strong positive control is usually more instructive and relevant to the hits of interest than a very or extremely strong positive control in RNAi HTS assays, and (ii) it has strongly been recommended that HTS assays incorporate as many controls as possible (http://nsrb. med.harvard.edu/assaydev.html), and different controls have different effect sizes. Controls in compound screens are straightforward, whereas siRNA controls in RNAi screens are usually neither straightforward nor as strong as the positive controls in compound HTS. For example, some RNAi screens in which cell viability is measured only use the background wells as a positive control, whereas others may use a weak positive control. Applying the same QC criterion to these two positive controls will lead to misleading QC results for detecting siRNAs with moderate or strong effects (i.e., judging the screens with very strong positive controls as good-quality assays even if they have poor quality and judging the screens with weaker positive controls as poor-quality assays even if they have good quality for detecting hits of interest that have moderate or strong effects).

4.4 Adoption of Effective Plate Designs

The results of QC assessment are affected not only by the choice of quality assessment metrics and their associated criteria, but also by the adoption of plate designs. Without the adoption of a suitable plate design, an assay with poor quality may be judged as having good quality, even using the best QC metric.

For example, suppose we conduct a primary screen with a very strong positive control and a negative control. In this screen, we use three different plate designs (i.e., designs A, B, and C as shown in Figure 2.1 in Chapter 2 in three plates). For these three plates, we evaluate data quality using both SSMD-based and Z-factor-based criteria on the positive and negative controls. The QC results in plate 1 are shown in Figure 4.4A1, which clearly indicates that the quality in plate 1 is good (using the SSMD-based criterion IIc in Table 4.2) or doable (using the Z-factor-based criterion in Table 4.3). However, if we check the raw data, as in Figure 4.4A2, there is a strong row effect. Because this is a primary screen, we can use the sample wells to adjust for the row effect, as described in Chapter 3. After adjusting for the row effect to obtain the data in Figure 4.4A3, the corresponding QC results actually indicate that the quality in plate 1 is inferior (using the SSMD-based criterion IIc in Table 4.2) or "screening essentially impossible" (using the Z-factor-based criteria in Table 4.3) (Figure 4.4A4). By contrast, when we adopt two more effective plate designs (i.e., designs B and C as shown in Figure 2.1 of Chapter 2) in plates 2 and 3, respectively, the QC results based on either raw or adjusted data suggest that the assay quality in plate 2 is poor (i.e., SSMD > -2) or "screening essentially impossible" (i.e., Z-factor < 0) (Figures 4.4B1 through B4 and 4.4C1 through C4).

In addition, by adopting an effective plate design in the very beginning of an experiment, we may save an assay that has a good quality but might have otherwise been judged as having poor quality. In a primary screen with a very strong positive control and a negative control, we obtained QC results as poor (using SSMD) or "screening essentially impossible" (using Z-factor) in most plates. The QC results in one typical plate are shown in Figure 4.5A. The image plot indicates there is a strong row effect (Figure 4.5B). Because the plate design adopted in this screen is an effective design (i.e., design C in Figure 2.1 in Chapter 2), we can use the negative control wells to adjust for the row effect. After adjusting for the row effect using the methods described in Chapter 3, we obtained data shown in Figure 4.5C. Based on the adjusted data, the QC results indicate that the quality is good (using SSMD) or "doable" (using Z-factor) (Figure 4.5D).

In confirmatory screens, a correlation plot between two replicate plates from the same source plate is a commonly used method to check data quality. Without the

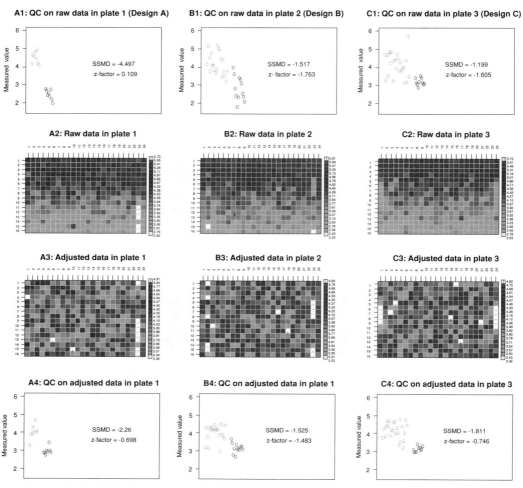

Figure 4.4   (See color insert following page 110.) Impact of plate design on quality assessment. Three designs (A, B, and C as shown in Figure 2.1 of Chapter 2) are adopted in plates 1, 2, and 3, respectively, in a primary screen with strong row effects. The red and green points respectively denote a very strong positive control and a negative control in A1, A4, B1, B4, C1, and C4.

adoption of effective plate designs, correlation analysis may also be misleading and useless. Suppose we have a confirmatory screen with a very strong positive control, a moderate positive control, and a negative control. In this confirmatory screen, plates 1 and 2 come from a source plate with plate design A (shown in Figure 2.1 in Chapter 2) and plates 3 and 4 come from another source plate with plate design C. The correlation plots for plates 1 and 2 as well as for plates 3 and 4 are shown in Figure 4.6.

The correlation between the raw values in plates 1 and 2 is 0.814, with a p-value of nearly 0 (Figure 4.6A1). Therefore, one may conclude that the reproducibility is very high and the assay quality very good. However, the image plot reveals a strong

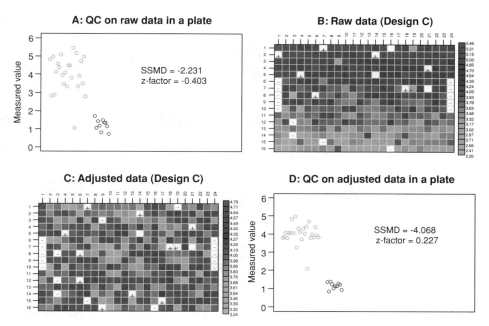

Figure 4.5   (See color insert following page 110.) An effective plate design saving an assay that might have been judged as poor quality. The screen had strong row effects and adopted design C as shown in Figure 2.1 of Chapter 2. The red and green points respectively denote a very strong positive control and a negative control in A and D.

bowl-shaped spatial effect in both plates (Figure 4.6A2 and A3). If the spatial effect is caused by the location of true hits, the high correlation may indicate good quality. On the other hand, if the spatial effect is caused by systematic experimental error, then the high correlation cannot indicate good quality at all. Unfortunately, because this source plate adopts design A, in which all negative control wells are arranged in column 23, we cannot tell whether this spatial effect is caused by the arrangement of true hits or by systematic experimental errors. In this design, we cannot use the negative control to adjust for the spatial effect.

By contrast, plates 3 and 4 come from a source plate with design C, in which the negative control wells are arranged across the plate in a balanced manner. The correlation plot of the raw values in plates 3 and 4 also indicates a correlation of 0.821, with a p-value of nearly zero. The image plot also reveals a strong bowl-shaped spatial effect in both plates (Figure 4.6B2 and B3). However, in this effective design, the values of the negative controls also have a bowl-shaped effect (Figure 4.6B2 and B3), which indicates that the spatial effect is caused by systematic errors and should be adjusted. As described in Section 3.3.3 of Chapter 3, using the negative control wells, we can adjust for the spatial effect to obtain adjusted data for plates 3 and 4, as shown in Figure 4.6B5 and B6. The image patterns in plates 3 and 4 are now very different (Figure 4.6B5 and B6), and the correlation between these two plates is −0.02 with a p-value of 0.72. Therefore, blindly using correlation analysis may

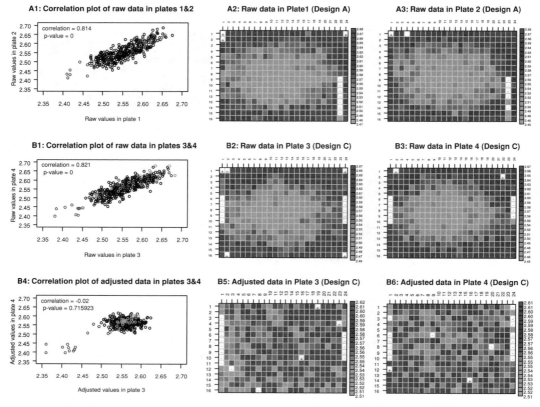

Figure 4.6   (See color insert following page 110.) Impact of systematic spatial effects on quality assessment using correlation between two replicate plates from the same source plate. The red, blue, green, and black points respectively denote a very strong positive control, a moderate positive control, a negative control, and sample siRNA wells in A1, B1, and B4.

lead to misleading results, and effective plate designs can help us make better use of correlation analysis.

4.5 Integration of Experimental and Analytic Approaches to Improve Data Quality

As we mentioned in the chapter introduction, the keys to integrating experimental and analytic approaches to improve data quality are the guidelines for selecting effective biological controls, the construction and usage of effective plate designs, and the development and adoption of effective analytic QC metrics. Table 4.5 provides a strategy for implementing both experimental and analytic approaches to improve data quality in RNAi HTS experiments.

Following the strategy in Table 4.5, we may select effective biological controls, display systematic errors, and then adjust for the errors when we can adopt effective plate designs in the very beginning. In a situation in which we may not control the

Table 4.5. A strategy for implementing experimental and analytic approaches to improve data quality in RNAi HTS experiments

1) Use the guidelines provided in Section 4.1 to select effective positive and negative controls.
2) Whenever possible, adopt effective plate design in the very beginning of the experiment following the guidelines described in Chapter 2.
3) Check for systematic errors of measurement using a negative reference:
    a) If systematic errors exist, try to adjust for them (see Section 3.3.3 of Chapter 3).
    b) If obvious systematic errors exist and cannot be adjusted, QC results reached using any analytic metric may not be reliable.
4) Adopt SSMD-based criteria for QC if systematic errors do not exist or have been adjusted.
5) Examine the plates failing QC to investigate potential causes.
    a) Check whether the failure is only caused by the contaminated positive controls:
        i) If yes, the plate may be used for hit selection.
        ii) Otherwise, redo the plate when possible.
    b) Check whether the failure is only caused by one or two extreme outliers in the controls:
        i) If yes and if the plate passes QC after removing the outliers, the plate may be used for hit selection.
        ii) Otherwise, redo the plate when possible.

plate design, this strategy may still help us to identify plates with bad quality and redo them or exclude them for further data analysis. In either situation, following this strategy will help us to obtain high-quality data. This strategy should be generally applicable to any assay in which the end point is a difference in signal compared with a reference sample, including enzyme, receptor, and cellular function assays, in addition to RNAi-based high-throughput screens.

4.6 Application

For the HCV primary siRNA screen described in Section 1.4 of Chapter 1, the biological controls are carefully selected as follows. a number of negative controls were purchased from various siRNA vendors and tested in the replicon system to guard against toxicity and off-target–mediated inhibition of HCV replication. The negative control used in the experiment was a nonsilencing siRNA purchased from Dharmacon. Positive controls were identified by testing siRNAs described to have efficacy against HCV replication in different replicon models of HCV replication, as well as by testing siRNAs targeting host factors with published links to HCV replication. Two positive control siRNAs were used: a very strong one that targeted the HCV replicon and was uniformly effective at knocking down HCV replication by 90% to 95% [121] and a weaker one (i.e., fairly strong one) that targeted hVAP33 [61;97]. In this instance, the weaker positive control was more instructive than the strong positive control, because the effectiveness of this control was more similar to

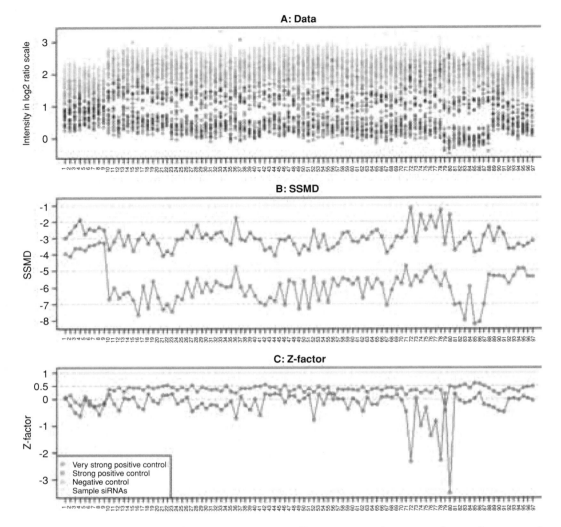

Figure 4.7   (See color insert following page 110.) Quality assessment in an HCV siRNA primary experiment. The x-axis in each panel denotes plate numbers. A point denotes the measured intensity in a well of a plate in A, a value of SSMD in a plate in B, and a value of Z-factor in a plate in C. The well types are denoted using different colors, as shown in the legend of C.

HTS hits than the siRNA targeting HCV itself. The control and sample wells were arranged as in Figure 2.1A in Chapter 2.

Using the SSMD-based criteria [166] shown in Table 4.2, by the very strong positive control (applying QC criterion IIc), 88 plates were excellent and 9 were good; by the strong (weaker) positive control (applying QC criterion IIb), 53 plates were excellent, 37 were good, and 7 were inferior (Figure 4.7B). By looking at the results for the two positive controls together, 90 plates were excellent or good by both controls (Table 4.6A). These plates all passed QC based on strategy 2 of

Table 4.6. Quality assessment using SSMD and *Z*-factor based on two positive controls in the HCV primary screen

A. Using SSMD						
		Strong Positive Control				
		Excellent	Good	Inferior	Poor	Total
Very	Excellent	**52**	**30**	6	0	88
Strong	Good	1	7	1	0	9
Positive	Inferior	0	0	0	0	0
Control	Poor	0	0	0	0	0
	Total	53	37	7	0	97

B. Using *Z*-factor							
		Strong Positive Control					
		Ideal	Excellent	Doable	Yes/No	SEI	Total
Very	Ideal	0	0	0	0	0	0
Strong	Excellent	0	0	**5**	0	3	8
Positive	Doable	0	0	**29**	0	52	81
Control	Yes/no	0	0	0	0	0	0
	SEI	0	0	2	0	6	8
	Total	0	0	36	0	61	97

Note: SEI, Screening essentially impossible.

Table 4.4. Only seven plates (i.e., 7/97 = 7%; plates 4, 36, 72, 74, 76, 78, and 80) were judged as inferior by the strong (weaker) positive control but good or excellent by the very strong positive control. The raw data displayed in Figure 4.7A also show that the strong positive control and the negative control do not differentiate well in these 7 plates but differentiate quite well in the remaining 90 plates. Therefore, the QC results using SSMD (Figure 4.7B) match with data (Figure 4.7A).

By contrast, if we use the *Z*-factor-based QC criterion in the HCV primary screen, we would obtain the following results: by the very strong positive control, 8 plates were "excellent," 81 were "doable," and 8 were "screening essentially impossible"; by the strong positive control, 36 plates were "doable" and 61 were "screening essentially impossible" (Figure 4.7C). When the results for the two positive controls were studied together, 55 plates were judged as "screening essentially impossible" by the strong positive control but as "doable" or "excellent" by the very strong positive control; 2 plates were judged as "screening essentially impossible" by the very strong positive control but as "doable" by the strong positive control (Table 4.6B). Therefore, it was difficult to evaluate the quality for each plate using *Z*-factor, especially for the 57 plates (i.e., 57/97 = 59%) with highly inconsistent QC results by the two positive controls. For example, plate 83 is judged as "excellent" if using the very strong positive control but as "screening essentially impossible" if using the strong positive control, which makes it hard to judge whether plate 83 should have "excellent" or "screening essentially impossible" quality when adopting the Z-factor-based QC criteria.

Therefore, the SSMD-based criteria produce much more consistent QC results than the Z-factor-based criterion: 93% of plates are good or excellent by both positive controls using SSMD-based criteria, whereas only 31% plates are "doable," "excellent", or "ideal" by both positive controls (Table 4.6). The judgments using SSMD are much more reasonable and match with the data better than those using Z-factor.

4.7 Discussion and Conclusions

This chapter provides a strategy for implementing both experimental and analytic approaches to improve data quality in RNAi HTS experiments (Table 4.5). The keys to this strategy are selection of effective biological controls, construction and use of effective plate designs, and development and adoption of effective analytic QC metrics.

Effective positive and negative controls are the means of gauging the size of siRNA effects. Hence the selection of effective positive and negative controls is critical for experiments to select siRNAs with a desired size of inhibition/activation effects. The guidelines for designs of controls provided in Chapter 2 should help to select effective positive and negative controls in the design of a genome-scale RNAi screen [166].

Whether in a primary screen or a confirmatory screen, a clear distinction between a positive control and a negative reference such as a negative control is an index for good quality. Many quality assessment measures, including signal-to-background ratio, signal-to-noise ratio, signal window, assay variability ratio, Z-factor, and SSMD, have been used to measure the degree of differentiation between a positive control and a negative reference [17;39;77;99;116;123;148;150;159;162;180;185]. The theoretically sound QC criteria are the SSMD-based criteria listed in Table 4.2. In the early stages of HTS research, the Z-factor-based criteria listed in Table 4.3 were commonly used. The disadvantage of the Z-factor-based criteria is that they do not take into account the different strength of different positive controls and thus may have issues, especially in experiments with more than two positive controls in the same direction. In such cases, it is better to use the SSMD-based criteria, even though these are more complicated. The best way to handle the complexity is to follow the guidelines listed in Table 4.2. A simple strategy for using SSMD-based criteria is to use criterion Ic for an up-regulated positive control and criterion IIc for a down-regulated positive control, unless we know the strength of a positive control or we have two positive controls in the same direction in an experiment.

Section 4.4 demonstrates that the adoption of effective plate designs as presented in Chapter 2 may have a huge impact on quality assessment results, regardless of whether we use SSMD and Z-factor in primary and confirmatory screens or correlation analysis among replicates in confirmatory screens. The blind use of differentiation metrics and correlation analysis may lead to misleading results. Effective plate designs are usually needed to help to reduce these misleading results.

The application in Section 4.6 demonstrates the importance and feasibility of integrating the above experimental and analytic approaches to improving data quality in genome-scale RNAi screens.

Hit Selection in Genome-Scale RNAi Screens without Replicates

5.1 Introduction

In an RNAi HTS, one primary goal is to select siRNAs with a desired size of inhibition or activation effect. The size of the siRNA effect is represented by the magnitude of difference between a tested siRNA and a negative reference group with no specific inhibition/activation effects. An siRNA with a desired size of effects in an HTS screen is called a hit. The process of selecting hits is called hit selection. There are two main strategies of selecting hits with large effects [161;174]. One is to use certain metric(s) to rank and/or classify the siRNAs by their effects and then to select the largest number of potent siRNAs that is practical for validation assays. The other strategy is to test whether an siRNA has effects strong enough to reach a pre-set level. In this strategy, false-negative rates (FNRs) and/or false-positive rates (FPRs) must be controlled.

As described in Chapter 1, a typical RNAi HTS project currently starts with a primary screen of single or pooled siRNAs, most of which have no replicate, and follows with one or more confirmatory screens in which each siRNA or pool has replicates. With the development of the platform of 1,536-well plates, more and more primary screens can also have replicates. The analytic methods for hit selection in screens without replicates differ from those with replicates. Therefore, I discuss these methods separately: without replicates in this chapter and with replicates in Chapter 6.

In this chapter, I describe various metrics; in Section 5.2, I discuss how to select a metric for hit selection in genome-scale RNAi screens without replicates. After choosing a metric, a decision rule must be constructed to judge whether an siRNA is a hit or non-hit, and this process is presented in Section 5.3. In the experimental design stage of a genome-scale RNAi project, we need to determine a suitable sample size for the achievement of reasonable FPRs and FNRs; thus I explore sample size determination in screens without replicates in Section 5.4. Finally, I demonstrate how to apply the described methods for selecting hits in real genome-scale RNAi screens without replicates in Section 5.5; conclusions are discussed in Section 5.6.

5.2 Methods for Hit Selection in Primary Screens without Replicates

There are many metrics used for hit selection in primary screens without replicates. The easily interpretable ones are fold change, mean difference, percent inhibition, and percent activity. However, the drawback common to all of these metrics is that they do not capture data variability effectively. To address this issue, researchers then turned to the z-score method or mean $\pm$ a standard deviation (SD) method, which can capture data variability in negative references. However, outliers are common in RNAi HTS experiments, and methods such as z-score are sensitive to outliers and can be problematic. Consequently, robust methods such as the z^*-score method (median $\pm$ a MAD method), B-score method, and quantile-based method have been proposed and adopted for hit selection [28;179;180]. For hit selection in RNAi screens, the major interest is the size of effect in a tested siRNA, which is represented by the magnitude of the difference between the siRNA and a negative reference group with no specific inhibition or activation effects. All the methods previously described attempt to estimate and test means of differences. They are not designed for assessing the magnitude of differences. SSMD can be used to assess the size of siRNA effects [161;167;182]. Therefore, SSMD should work best for hit selection in genome-scale RNAi screens.

5.2.1 A Negative Reference to Represent siRNAs with No Specific Effects

All the methods described in this section require a negative reference group to represent the siRNAs with no specific inhibition/activation effects. Thus the choice of a negative reference is crucial in RNAi screens. Two common choices for the negative reference are (i) negative control wells and (ii) sample wells.

When an effective negative control is available in a screen and the number of wells per plate for the negative control is large (>10), the negative control is an ideal choice. However, in many screens, the negative control may not work effectively to represent siRNAs with no specific effects, or the number of wells per plate for the negative control is too small (<6). In such a case, it is a good idea to use the majority of sample wells as the negative reference for the primary screen because the majority of sample siRNAs in a primary screen's plate tend to have no or extremely weak specific inhibition/activation effects and because the number of sample wells per plate is large (approximately 300 in a 384-well plate). The use of the majority of sample wells as a negative reference may lead to more robust and more stable results because of the large sample size. In some cases, there is no effective negative control available at all. In such cases, we may only use the sample wells as a negative reference in a primary screen. In a confirmatory screen, the siRNAs arranged in a plate are selected from a primary screen or from other studies and are supposed to have specific inhibition or activation effects. Thus, in most confirmatory screens, we may only use a negative control as a negative reference for hit selection.

5.2.2 Plate-Wise versus Experiment-Wise

In the process of hit selection, we usually face the following question: Should we perform these analyses on a plate-by-plate basis (called plate-wise) or on all the plates in an experiment (called experiment-wise)?

If there are different systematic errors in different plates (e.g., different plates may have different transfection efficiencies and hence have different centers and variability), plate-wise analysis can adjust for the different systematic errors in different plates, whereas experiment-wise analysis cannot. On the other hand, there is the possibility that a cluster of active siRNAs will be located within a single plate, which will cause the variability of sample wells in this plate to be inflated. The plate-wise sample-based method may not detect these true hits. If an effective negative control is used as a negative reference and the number of wells for the negative control in each plate is small (e.g., < 8 wells/plate), then experiment-wise analysis can substantially increase the power in estimating the mean and SD of the negative reference. The existence of enriched plates is not very common in a primary RNAi screen. Therefore, plate-wise analysis is more commonly used, especially when the sample wells are used as a negative reference.

5.2.3 Hit Selection Metrics and Their Calculation

Metrics and their calculation for hit selection in primary screens without replicates are listed in Table 5.1.

z-score. The *z*-score method is equivalent to the mean $\pm a$ SD method, in which the siRNAs with bigger measured value than the mean $+ a$ SD or smaller than the mean $- a$ SD are selected as hits, where a is a pre-set constant usually being 2 or 3. This method relies on the *z*-score of the standard normal distribution $N(0,1)$ and is thus also called the *z*-score method. The *z*-score method addresses the question of what would happen if an investigated siRNA truly comes from the negative reference population with no specific inhibition or activation effects. a is also the cutoff (or critical value) of *z*-score.

z*-score. In practice, the measured values are not normally distributed. Long-tailed or even skewed distributions may occur. Outliers appear frequently in RNAi HTS data. True hits should behave differently from the siRNAs, which do not have specific silencing effects. Because the majority of siRNAs in a primary HTS experiment do not have specific silencing effects, true hits (especially strong ones) should behave like outliers. To obtain estimates for the center and variability similar to mean and SD, but more robust to outliers and the violation of normal assumption, one

Table 5.1. Metrics and their calculation for hit selection in primary screens without replicates

Hit Selection Metric	Calculation Formula	
	Regular Version	Robust Version
Fold change	**A1:** $\dfrac{Y_{siRNA}}{\bar{Y}_-}$	**A2:** $\dfrac{Y_{siRNA}}{\tilde{Y}_-}$
Percent activity	**B1:** $\dfrac{Y_{siRNA}}{\bar{Y}_+} \times 100$	**B2:** $\dfrac{Y_{siRNA}}{\tilde{Y}_+} \times 100$
Percent inhibition	**C1:** $\dfrac{Y_{siRNA} - \bar{Y}_-}{\bar{Y}_+ - \bar{Y}_-} \times 100$	**C2:** $\dfrac{Y_{siRNA} - \tilde{Y}_-}{\tilde{Y}_+ - \tilde{Y}_-} \times 100$
z-score	**D1:** $\dfrac{Y_{siRNA} - \bar{Y}_-}{SD_-}$	**D2:** $\dfrac{Y_{siRNA} - \tilde{Y}_-}{MAD_-}$, called z^*-score
SSMD	**E1:** $\dfrac{Y_{siRNA} - \bar{Y}_-}{\sqrt{2}SD_-}$	**E2:** $\dfrac{Y_{siRNA} - \tilde{Y}_-}{\sqrt{2}MAD_-}$
B-score method	*B*-score [20] is a robust metric similar to z^*-score but with an additional adjustment for positional effects.	
Quantile method	The lower boundary for hit selection is the smallest observed value greater than $Q_1 - c$ interquartile range (IQR) and the upper boundary for hit selection is the biggest observed value smaller than $Q_3 + c$ IQR, where IQR $= Q_3 - Q_1$, Q_1 and Q_3 are the 1st and 3rd quartiles of measured values in a negative reference, respectively, and c is a pre-set constant [180].	
Bayesian method	Adopts robust estimation in the prior and controls false discovery rate via a direct posterior approach [176].	

Y_{siRNA} denotes a measured value usually in log-scale of an investigated siRNA; $\bar{Y}_-$, $\tilde{Y}_-$, SD_-, and MAD_- denote the mean, median, standard deviation (SD), and median of absolute deviation (MAD) of measured values in a negative reference, respectively.

common choice is median and MAD [20;28;108;180]. MAD represents the median of absolute deviations, that is:

$$MAD = 1.4826 \, \text{median}(|y_i - \text{median}(y)|).$$

The constant of 1.4826 is chosen so that MAD is equivalent to SD when the measured values are normally distributed. Similar to mean $\pm$ a SD, the method based on median and MAD is median $\pm$ a MAD, where a is often set to be 2 or 3. Similar to z-score, we have z^*-score as Formula D2 in Table 5.1.

B-score. *B*-score [20] is a robust metric similar to z^*-score but with an additional adjustment for positional effects. Technically, the *B*-score method is equivalent to doing the following two steps at once: (i) applying smoothing to adjust for positional effect during the normalization stage and (ii) calculating the z^*-score using sample wells as the negative reference based on the normalized data. The *B*-score method adjusts for systematic positional effects regardless of whether the adjustment is needed. When these positional effects are caused by measurement error, the adjustment is appropriate. However, when these positional effects are caused by the clusters of hits, the adjustment may lead to missing the cluster of hits. This can be problematic if one or more clusters of true hits are placed in a small area of a plate. Thus the best strategy is to conduct the following analyses step by step: (i) apply the methods described in Chapter 3 to determine whether there are any systematic positional effects; (ii) if there are, investigate whether they are caused by systematic experimental errors; (iii) if the systematic positional effects are caused by systematic experimental errors, adopt necessary and suitable adjustment/normalization methods as described in Chapters 2 and 3 to adjust for the identified systematic errors; and finally, (iv) apply a suitable analytic method (chosen from not only z^*-score method, but also other alternatives described in Table 5.1) to the normalized data for hit selection.

SSMD. SSMD is the mean difference penalized by the inconsistency (i.e., variability) of the difference between an siRNA and a negative reference. In screens without replicate, there is a linear relationship between z-score and SSMD, namely z-score $= \sqrt{2}$ SSMD, when the estimation in Table 5.1 is used. Note that this relationship does not exist for screens with replicates. An advantage is that we can use SSMD to classify the size of siRNA effects as in Table 5.2 when the true value of SSMD is known. See Chapter 8 for more details about how the cutoffs in this table were derived. Another advantage is that z-score cannot be applied in screens with replicates, whereas SSMD can.

The usual process for calculating SSMD listed in Table 5.1 comes from the method-of-moments estimation, which is a biased estimate. A uniformly minimal variance unbiased estimate of SSMD is:

$$\hat{\beta} = \frac{Y_{\text{siRNA}} - \bar{Y}_-}{\sqrt{\frac{2}{K}(n_- - 1)}\ \text{SD}_-} \tag{5.1}$$

where $K \approx n_- - 2.48$ and n_- is the sample size in the negative reference. See Chapter 8 for more details regarding how this estimate is derived.

Outliers. To reduce the impact of outliers, one approach is to adopt the robust version of hit selection metrics as in Formulas A2, B2, C2, D2, and E2 in Table 5.1. Another approach is to calculate $\bar{Y}_-$ and SD$_-$ after excluding outliers in the negative reference. This approach can be conducted plate-wise or experiment-wise. For the plate-wise calculation, in the plate in which the siRNA is placed, boxplot parameters are used to identify outliers in a negative reference group. n_-, $\bar{Y}_-$, and SD$_-$ are respectively the sample size, mean, and standard deviation of measured values (usually after log-transformation) in the negative reference, excluding the identified outliers in that plate. This calculation is often adopted in most screens, especially when the sample wells are used as a negative reference. We usually adopt

Table 5.2. The size of siRNA effects classified using the true value of SSMD

Effect Direction	Effect Subtype	SSMD Threshold
Up-regulated (i.e., increasing activity)	Extremely strong	$SSMD \geq 5$
	Very strong	$5 > SSMD \geq 3$
	Strong	$3 > SSMD \geq 2$
	Fairly strong	$2 > SSMD \geq 1.645$
	Moderate	$1.645 > SSMD \geq 1.28$
	Fairly moderate	$1.28 > SSMD \geq 1$
	Fairly weak	$1 > SSMD \geq 0.75$
	Weak	$0.75 > SSMD > 0.5$
	Very weak	$0.5 \geq SSMD > 0.25$
	Extremely weak	$0.25 \geq SSMD > 0$
Zero	Zero	$SSMD = 0$
Down-regulated (i.e., decreasing activity)	Extremely weak	$-0.25 \leq SSMD < 0$
	Very weak	$-0.5 \leq SSMD < -0.25$
	Weak	$-0.75 < SSMD < -0.5$
	Fairly weak	$-1 < SSMD \leq -0.75$
	Fairly moderate	$-1.28 < SSMD \leq -1$
	Moderate	$-1.645 < SSMD \leq -1.28$
	Fairly strong	$-2 < SSMD \leq -1.645$
	Strong	$-3 < SSMD \leq -2$
	Very strong	$-5 < SSMD \leq -3$
	Extremely strong	$SSMD \leq -5$

this calculation to obtain the SSMD value, especially when Formula 5.1 is used. For the experiment-wise calculation, boxplot parameters are first used to identify outliers in a negative reference group across all plates in the experiment. n_-, $\bar{Y}_-$, and SD_- are respectively the sample size, mean, and standard deviation of measured values in the negative reference, excluding the identified outliers in the whole experiment. This calculation may be used when a negative control with a small sample size is used as a negative reference. This calculation is implemented in the prior in the Bayesian methods described in Zhang et al. [176].

5.2.4 Comparison of Metrics for Hit Selection Using Example Data Sets

Here we use the example data sets in Table 5.3 to demonstrate the use of various metrics for hit selection in primary screens without replicates. The calculated values for various metrics based on these data sets are shown Figure 5.1. The only difference between data sets A and B is that there exist two outliers in data set A, one in the positive control and one in the negative control. As demonstrated in Figure 5.1A and B, the median-based fold change (i.e., fold change2) is 1.9 for both data sets A and B, whereas the mean-based fold change (i.e., fold change1) is 1.4 for data set A

Table 5.3. Example data sets for demonstrating the usage of metrics for hit selection in RNAi primary screens without replicates (focusing on one siRNA only in each data set)

	Data Set A	Data Set B	Data Set C
siRNA	6.0	6.0	6.0
Negative control	2.1, 2.2, 2.4, 2.7, 3, 3.2, 3.4, 3.5, 3.6, 16	2.1, 2.2, 2.4, 2.7, 3, 3.2, 3.4, 3.5, 3.6, 3.9	0.1, 0.4, 0.9, 1.2, 1.9, 2.7, 3.4, 5.2, 6.2, 7.8
Positive control	6.0, 6.2, 6.4, 16.6	6.0, 6.2, 6.4, 6.6	4.0, 5.2, 7.4, 8.6

but 2 for data set B, which indicates that the median-based fold change is robust to outliers, whereas the mean-based fold change is sensitive to outliers. Similar results are obtained for percent inhibition and percent activity.

The value in the siRNA is consistently above all the values in the negative control in Figure 5.1B, whereas the value in the siRNA is less than at least two values in the negative control in Figure 5.1C, which indicates that the siRNA separates well from the negative control in Figure 5.1B but not in Figure 5.1C. However, the mean-based fold change has the same value in both Figure 5.1B and C, and the median-based fold change is smaller in Figure 5.1B than in Figure 5.1C, which indicates that the siRNA separates from the negative control in Figure 5.1B no better than in Figure 5.1C. Thus both mean-based and median-based fold changes produce misleading results. Percent inhibition and percent activity give similar misleading results (Figure 5.1). The issues with fold change and percent inhibition are further demonstrated in real RNAi screens in Section 5.5. These issues can be addressed using methods that incorporate information about data variability, as described in following subsections.

z-score. The z-score method takes into account data variability by penalizing data variability in the negative reference. In Figure 5.1, the z-score is 4.8 in B and 1.1 in C, which correctly indicates the separation of the siRNA from the negative control. The z-score method is based on the assumption of a normal distribution of data and is very sensitive to outliers. As demonstrated in Figure 5.1, one outlier in the negative reference may dramatically change the z-score value from 4.8 (B) to 0.4 (A).

z^*-score. The z^*-score is 4.3 for the siRNA in data set B and 1.5 for the siRNA in data set C. Thus, like z-score, z^*-score takes into account data variability, and its values correctly indicate the separation of the siRNA from the negative control. In addition, z^*-score is 4.3 for both siRNAs in data sets A and B, demonstrating that z^*-score is robust to outliers.

SSMD. Calculated based on Formula 5.1 after excluding outliers using boxplot statistics, the value of SSMD is 3.1 for the siRNA in data set B and 0.7 for the siRNA in data set C (Figure 5.1). Thus, like z-score and z^*-score, SSMD takes into account data variability, and its values correctly indicate the separation of the siRNA from the negative control. The values of SSMD are 3.5 for the siRNA in data set A and 3.1 for the siRNA in data set B, demonstrating that SSMD in Formula 5.1 is robust to

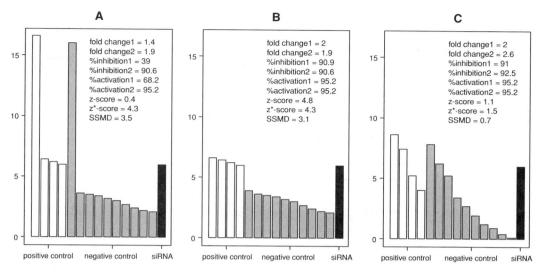

Figure 5.1   Examples of 4 wells for a positive control (white bars), 10 wells for a negative control (grey bars), and 1 well for an siRNA (black bar) in three data sets shown in Table 5.3, demonstrating the use of fold change, percent inhibition, percent activation, z-score, z^*-score, and SSMD in RNAi screens without replicates. The data in the positive control, negative control, and an siRNA in (A) are the same as the corresponding data in (B) except for one outlier in the positive control and one outlier in the negative control in (A). Data variability in (C) is larger than in (B). Fold change1, %inhibition1, and %activation1 are all based on means of positive and/or negative control; fold change2, %inhibition2, and %activation2 are all based on medians of positive and/or negative control.

outliers. In addition, applying the criteria listed in Table 5.2, we can roughly classify the size of effects as very strong for the siRNAs in data sets A and B and weak for the siRNA in data set C.

5.3 Decision Rules for Hit Selection in RNAi Screens

For hit selection in RNAi screens, the key is to find an analytic metric to effectively quantify knockdown effects of siRNAs and then to construct a decision rule based on this metric to identify siRNAs with large effects on a biological response of interest. Sections 5.2 presents various metrics for quantifying siRNA effects. Some of these metrics are estimates of population parameters, such as fold change, percent activity, and SSMD, whereas others are testing statistics such as z-score and z^*-score. The testing statistics aim at testing the null hypothesis about a parameter, such as mean difference. A parameter has a true value (or, more accurately, a population value) in a distributional level. We usually do not know the true value of a parameter. However, we can estimate and test it based on some measured values (or, in statistical terms, some random samples). The estimated value of a parameter based on random samples may deviate from its true value due to the stochastic features of the measured response. Therefore, to determine the decision rule for hit selection, we need to control false positives and false negatives based on the estimated values of mean difference or SSMD.

False positives and false negatives are defined by a statistical parameter. Many traditional methods, including z-score method or t-test for hit selection, use mean difference as a parameter to define false positives and false negatives. By contrast, the SSMD method uses SSMD as a parameter to define false positives and false negatives. Consequently, the decision rules in the z-score method differ from those in the SSMD method. Therefore, I introduce the definition of false positives and false negatives in Section 5.3.1, describe decision rules in the z-score method in Section 5.3.2, and present decision rules in the SSMD method in Section 5.3.3.

5.3.1 Definition of False Positives and False Negatives

After choosing a parameter for hit selection, we need to set up a threshold for the true value of this parameter for defining a true hit. When mean difference is chosen as the parameter for hit selection, as in the z-score method, the following rule is commonly used for defining true hits in the down-regulated direction: the siRNAs with true values of mean difference less than zero are defined as true hits, and the siRNAs with true values of mean difference greater than or equal to zero are defined as true non-hits. Because the true value of the mean difference is unknown and we can only obtain its estimated value, we need to control false positives and false negatives, defined as follows: an siRNA that has a true value of mean difference greater than or equal to 0 but is declared as a hit based on its estimated value is a false positive; the siRNA that has true value of mean difference less than zero but is declared as a non-hit based on its estimated value is a false negative.

The definition of false positives and false negatives in the SSMD method differs from that in the z-score method because it captures not only the effect size of siRNAs in which we are interested, but also the effect size of siRNAs that we do not want to include in the selected list of hits. Consequently, in the z-score method, we only need to specify one value (usually zero) for the tested parameter (i.e., mean difference), whereas in the SSMD method, the definition of false positives and false negatives requires the specification of two different values, β_1 and β_2, for the tested parameter SSMD. Using the case of selecting siRNAs with very strong effects or even stronger effects and avoiding siRNAs with extremely weak effects or even weaker effects in the down-regulated direction as an example, an siRNA that has a true value of SSMD greater than or equal to -0.25 but is declared as a hit based on its estimated value is a false positive; an siRNA that has a true value of SSMD less than or equal to -3 but is declared as a non-hit based on its estimated value is a false negative. In this case, $\beta_1 = -3$ and $\beta_2 = -0.25$.

5.3.2 Decision Rules in the z-Score Method

When using the z-score method for hit selection in screens without replicates, the decision rule for selecting down-regulated hits is as follows.

Decision rule 5.1 (down-regulation). Declare an siRNA as a hit if it has a z-score value less than or equal to a critical value a and as a non-hit otherwise.

To use this decision rule, we need to determine the critical value a so that a suitable FPR can be achieved.

False-positive rate. This decision rule actually comes from testing the mean difference. Let μ denote mean difference. Then this decision rule corresponds to a test of the null hypothesis H_0: $\mu \geq 0$. The FPR in decision rule 5.1 is the probability that the z-score is less than or equal to a when the true value of the mean difference is greater than or equal to zero, namely, FPR $= \Pr(z\text{-score} \leq a|\mu \geq 0)$. Given $\mu \geq 0$, the maximal FPR is called the false-positive level (FPL), which is achieved at $\mu = 0$, namely, FPL $= \Pr(z\text{-score} \leq a|\mu = 0)$. Using normal approximation, the z-score has a standard normal distribution. Thus FPL $= \Phi(a)$, where Φ is the cumulative distribution function of the standard normal distribution. The FPL is applied to any siRNA. When we focus on one specific siRNA and treat the z-score z_{obs} of this siRNA as a critical value, then FPL becomes the p-value with respect to this observed value. That is, the p-value for an siRNA with an observed z-score value z_{obs} is $p\text{-value} = \Pr(z\text{-score} \leq z_{\text{obs}}|\mu = 0) = \Phi(z_{\text{obs}})$.

False-negative rate. Similarly, the FNR is the probability that the z-score is greater than a when the true value of mean difference is less than zero, namely, FNR $=$ $\Pr(z\text{-score} > a|\mu < 0)$. Given $\mu < 0$, the upper limit of FNR is called the false-negative level (FNL), which is achieved at $\mu = 0$, namely, FPL $= \Pr(z\text{-score} > a|\mu = 0)$. Under a normal assumption, FNL $= 1 - \Phi(a)$.

Traditionally, to find the critical value a, we control FPL to be α, where α equals 0.05 or 0.01; then $a = a_\alpha$ where a_α is a critical value such that $\Phi(a_\alpha) = \alpha$.

All the above are for the down-regulated direction. Similarly, the decision rule for selecting up-regulated hits is as follows.

Decision rule 5.2 (up-regulation). Declare an siRNA as a hit if it has a z-score value greater than or equal to a critical value a and as a non-hit otherwise.

Decision rules 5.1 and 5.2 and their associated FPRs, FNRs, and p-values are also listed in the left panel of Table 5.4.

5.3.3 SSMD-Based Decision Rules

For convenience, let β denote SSMD. When using the SSMD method for hit selection in screens without replicates, the decision rule for selecting down-regulated hits is as follows.

Decision rule 5.3 (down-regulation). Declare an siRNA as a hit if it has an estimated value $\hat{\beta}$ of SSMD less than or equal to a critical value β^* and as a non-hit otherwise.

To use this decision rule, we need to determine the critical value β^* so that a suitable FPR and FNR can be achieved.

Table 5.4. Decision rules, FPL, FNL, and p-value for hit selection in RNAi primary screens without replicates

Direction	Calculation Formula	
	z-Score Method	SSMD Method
Down-regulation	**Decision Rule 5.1**: Any siRNA is a hit if it has z-score $\leq a$ and a non-hit otherwise	**Decision Rule 5.3**: Any siRNA is a hit if estimated SSMD $\leq \beta^*$ and a non-hit otherwise
	Formula 5.1A: FPL $= \Phi(a)$	**Formula 5.3A**: FPL $= F_{t(v,b\beta_2)}\left(\dfrac{\beta^*}{k}\right)$
	5.1B: FNL $= 1 - \Phi(a)$	**5.3B**: FNL $= 1 - F_{t(v,b\beta_1)}\left(\dfrac{\beta^*}{k}\right)$
	5.1C: p-value $= \Phi(z_{\text{obs}})$	**5.3C**: p-value $= F_{t(v,b\beta_2)}\left(\dfrac{\beta_{\text{obs}}}{k}\right)$
Up-regulation	**Decision Rule 5.2**: Any siRNA is a hit if it has z-score $\geq a$ and a non-hit otherwise	**Decision Rule 5.4**: Any siRNA is a hit if estimated SSMD $\geq \beta^*$ and a non-hit otherwise
	Formula 5.2A: FPL $= 1 - \Phi(a)$	**Formula 5.4A**: FPL $= 1 - F_{t(v,b\beta_2)}\left(\dfrac{\beta^*}{k}\right)$
	5.2B: FNL $= \Phi(a)$	**5.4B**: FNL $= F_{t(v,b\beta_1)}\left(\dfrac{\beta^*}{k}\right)$
	5.2C: p-value $= 1 - \Phi(z_{\text{obs}})$	**5.4C**: p-value $= 1 - F_{t(v,b\beta_2)}\left(\dfrac{\beta_{\text{obs}}}{k}\right)$

Note: FNL, false-negative level; FPL, false-positive level. $\Phi(\cdot)$ and $F_{t(v,b\beta)}(\cdot)$ are the cumulative distribution functions of the standard normal distribution $N(0,1)$ and noncentral t-distribution $t(v, b\beta)$, respectively; $b = \sqrt{2}/\sqrt{1 + 1/n_-}$; $v = n_- - 1$; n_- is the sample size in the negative reference; $k = \sqrt{K}/\sqrt{2(n_-1)}.\sqrt{1 + 1/n_-}$ when SSMD is estimated using Formula 5.1 and $k = \sqrt{1 + 1/n_-}$ when SSMD is estimated using Formula E1 in Table 5.1. z_{obs} and β_{obs} are the z-score and estimated SSMD value of an siRNA, respectively. β_1 and β_2 are population values of SSMD to indicate large and small effects, respectively.

SSMD can effectively assess the size of siRNA effects and can classify them into different categories such as zero, extremely weak, very weak, and so on (as shown in Table 5.2). Based on SSMD, we can control the FPR not only with respect to no effect, but also with respect to extremely weak effects or even very weak effects. Similarly, we can control the FNR with respect to extremely strong, very strong, and strong. For example, in the down-regulated direction, we can use $\beta_1 = -3$ to indicate large effects and $\beta_2 = -0.25$ to indicate small or no effects. The corresponding FPR is the probability that an siRNA is selected as a hit based on estimated values given that the true value of SSMD is greater than or equal to β_2; the FNR is the probability that an siRNA is selected as a non-hit based on estimated values given that the true value of SSMD is less than or equal to β_1.

The FPR in decision rule 5.3 is the probability that the SSMD estimate is less than or equal to β^* when the true value of SSMD is greater than or equal to β_2, namely, FPR $= \Pr(\hat{\beta} \leq \beta^* | \beta \geq \beta_2)$. Its corresponding FPL is $\Pr(\hat{\beta} \leq \beta^* | \beta = \beta_2)$.

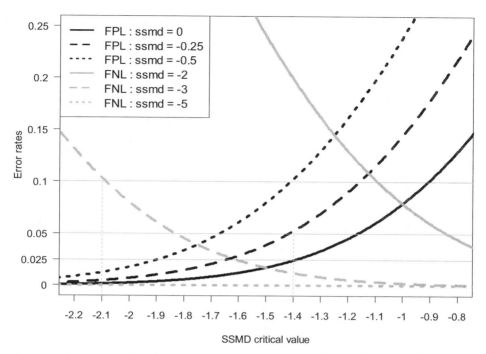

Figure 5.2   The changes of error rates (i.e., false-positive level [FPL] and false-negative level [FNL]) versus critical value when using the majority of sample wells as a negative reference for selecting down-regulated hits in a primary screen without replicates.

Similarly, the FNR in decision rule 5.3 is the probability that the SSMD estimate is greater than or equal to β^* when the true value of SSMD is less than or equal to β_1, namely, $FNR = Pr(\hat{\beta} \geq \beta^* | \beta \leq \beta_1)$. Its corresponding FPL is $Pr(\hat{\beta} \geq \beta^* | \beta = \beta_1)$. The formulas to calculate FPL, FNL, and p-value based on the noncentral distribution are Formulas 5.3A through 5.3C in Table 5.4.

Based on Formulas 5.3A, 5.3B, 5.4A, and 5.4B in Table 5.4, we can calculate the theoretical FPL and FNL corresponding to each set of values for β^*, β_1, and β_2. In a primary screen using 384-well plates, the majority of sample wells may be used as the negative reference in a plate. In such a case, n_- is approximately 300. To select down-regulated hits, we can use Formulas 5.3A and 5.3B in Table 5.4 to calculate theoretical FPLs with respect to $\beta_2 = 0, -0.25, -0.5$ and theoretical FNLs with respect to $\beta_1 = -2, -3, -5$. Figure 5.2 shows the calculated FPLs and FNLs when $n_- = 300$. The commonly used error rates are 0.05, 0.025, and 0.01 in one direction. From Figure 5.2, a critical value between -1.8 and -1.4 can control FPL with respect to $\beta_2 = 0$ to be less than 0.025, FPL with respect to $\beta_2 = -0.25$ to be less than 0.051, and FNL with respect to $\beta_1 = -3$ to be less than 0.05. A critical value between 1.9 and 2.1 can control FPL with respect to $\beta_2 = 0$ to be less than 0.005, FPL with respect to $\beta_2 = -0.25$ to be less than 0.01, and FNL with respect to $\beta_1 = -3$ to be less than 0.10. Therefore, any critical value between -2.1 and -1.4 for SSMD is theoretically reasonable and may maintain a balanced control of both FPR for including siRNAs with small or no down-regulated effects and FNR for excluding siRNAs with large down-regulated effects [165].

The choice of an exact critical value between -2.1 and -1.4 in a real experiment relies on the refined tolerance of false positives and false negatives and the capacity of follow-up studies after that experiment. For example, if one has a low tolerance in missing hits with SSMD less than -2 or -3, then one may choose a critical value between -1.6 and -1.4. On the other hand, if follow-up studies have a low capacity of including selected hits, then one may choose a critical value between -2.1 and -1.8.

Similarly, the decision for selecting up-regulated hits is as follows:

Decision rule 5.4 (up-regulation). Declare an siRNA as a hit if it has an estimated value $\hat{\beta}$ of SSMD greater than or equal to a critical value β^* and as a non-hit otherwise.

The formulas for calculating FPL, FNL, and p-value in this decision rule for selecting up-regulated hits are Formulas 5.4A through 5.4C in Table 5.4. See Chapter 8 for more details about how these formulas are derived.

The strategy for hit selection described previously is to determine a critical value of SSMD through various FPLs and FNLs so that we can obtain reasonable levels of FPRs and FNRs. Another strategy is to fix a pre-set FPL (or FNL) first and then to calculate the corresponding critical value of estimated SSMD and FNL (or FPL). When the FPL with respect to β_2 is pre-set to be α_1, the corresponding critical value is $\beta^* = \beta_{\alpha_1}$, where β_{α_1} is obtained by solving $F_{t(n_- - 1, b\beta_2)}(\beta_{\alpha_1}/k) = \alpha_1$. The corresponding FNL with respect to β_1 is FNL $= 1 - F_{t(n_- - 1, b\beta_1)}(\beta_{\alpha_1}/k)$. Similarly, we can pre-set FNL with respect to β_1 to be α_2 and calculate the corresponding critical value and FPL. The formulas are shown in Table 5.5.

5.4 Sample Size Determination

As described in Chapter 2, we need to determine a sample size for the achievement of certain FNLs and FPLs in the experimental design stage of a genome-scale RNAi project. We can use the formulas presented in Table 5.4 to calculate theFPRs and FNRs corresponding to various sample sizes. The critical issue of sample size determination in a primary screen without replicates is to determine the number of replicates (i.e., number of wells per plate) in the negative reference group. An essential consideration for hit selection in a primary screen is the capacity available for confirmation screening or other investigations after the primary screen. In a typical primary screen, there are approximately 20,000 siRNAs, and the major goal is to select 300 to 800 siRNAs in one direction for follow-up research. If we control the FPLs with respect to extremely weak or no effects to be 0.05, 0.025, and 0.01 for one direction, we would obtain 1,000, 500, and 200 hits, respectively, even if all the

Table 5.5. SSMD-based decision rules, FPL, and FNL in RNAi primary screens without replicates when a pre-set FPL or FNL is fixed

Direction	Calculation Formula	
	Fix FPL w.r.t. β_2 be α_1	Fix FNL w.r.t. β_1 to be α_2
Down-regulation	**Decision Rule 5.5**: Any siRNA is a hit if estimated SSMD $\leq \beta_{\alpha_1}$ and a non-hit otherwise, where β_{α_1} is obtained by solving $$F_{t(n_- - 1, b\beta_2)}\left(\frac{\beta_{\alpha_1}}{k}\right) = \alpha_1$$ **Formula A:** $$FNL = 1 - F_{t(n_- - 1, b\beta_1)}\left(\frac{\beta_{\alpha_1}}{k}\right)$$	**Decision Rule 5.6**: Any siRNA is a hit if estimated SSMD $\leq \beta_{\alpha_2}$ and a non-hit otherwise, where β_{α_2} is obtained by solving $$F_{t(n_- - 1, b\beta_1)}\left(\frac{\beta_{\alpha_2}}{k}\right) = 1 - \alpha_2$$ **Formula B:** $$FPL = F_{t(n_- - 1, b\beta_2)}\left(\frac{\beta_{\alpha_2}}{k}\right)$$
Up-regulation	**Decision Rule 5.7**: Any siRNA is a hit if estimated SSMD $\geq \beta_{\alpha_1}$ and a non-hit otherwise, where β_{α_1} is obtained by solving $$F_{t(n_- - 1, b\beta_2)}\left(\frac{\beta_{\alpha_1}}{k}\right) = 1 - \alpha_1$$ **Formula C:** $$FNL = F_{t(n_- - 1, b\beta_1)}\left(\frac{\beta_{\alpha_1}}{k}\right)$$	**Decision Rule 5.8**: Any siRNA is a hit if estimated SSMD $\geq \beta_{\alpha_2}$ and a non-hit otherwise, where β_{α_2} is obtained by solving $$F_{t(n_- - 1, b\beta_1)}\left(\frac{\beta_{\alpha_2}}{k}\right) = \alpha_2$$ **Formula D:** $$FPL = 1 - F_{t(n_- - 1, b\beta_2)}\left(\frac{\beta_{\alpha_2}}{k}\right)$$

Note: FNL, false-negative level; FPL, false-positive level; w.r.t., with respect to. $F_{t(n_- - 1, b\beta)}(\cdot)$ is the cumulative distribution function of a noncentral t-distribution $t(n_- - 1, b\beta)$; $b = \sqrt{2}/\sqrt{1 + 1/n_-}$; n_- is the sample size in the negative reference; $k = \sqrt{K}/\sqrt{2(n_- - 1)}.\sqrt{1 + 1/n_-}$ when SSMD is estimated using Formula 5.1 and $k = \sqrt{1 + 1/n_-}$ when SSMD is estimated using Formula E1 in Table 5.1. β_1 and β_2 are population values of SSMD to indicate large and small effects, respectively.

20,000 siRNAs have extremely weak or no effects. Clearly, an FPL of 0.05 is too large, an FPL of 0.025 might be acceptable, and a FPL of 0.01 is preferred.

Calculated using Formula A in Table 5.5, the FNLs (with respect to $\beta_1 = -1.28$, -1.645, -2, -3, and -5) under the control of FPL = 0.025 and 0.01 with respect to $\beta_2 = 0$ are shown in Figure 5.3A and B, respectively; the FNLs under the control of FPL = 0.025 and 0.01 with respect to $\beta_2 = -0.25$ are shown in Figure 5.3C and D, respectively. The figure shows that a choice of four or eight wells per plate for the negative control is not enough to achieve an acceptable FNRs; a choice of 16 wells per plate is acceptable, and a choice of 20 to 24 wells per plate is preferable [175].

The FNL is relatively large for 4 to 12 negative references in a plate and becomes fairly flat as sample size increases above 16 (or, in some cases, 20–24). For example, if the FPL with respect to $\beta_2 = 0$ is controlled to be 0.025 (Figure 5.3A), then the curves of FNLs with respect to $\beta_1 = -1.28$, -1.645, -2, -3 go down relatively quickly when the sample size increases from 4 to 16 and become relatively flat when

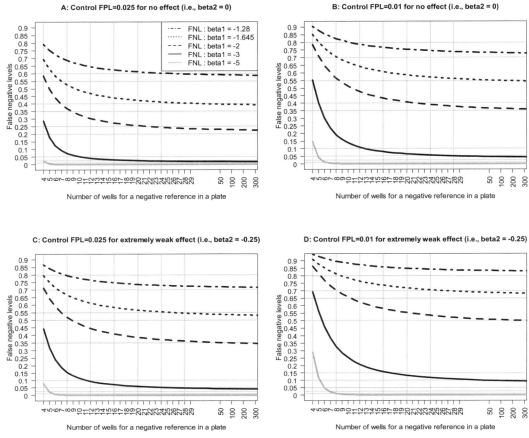

Figure 5.3   The changes of false-negative level (FNL) versus sample size by controlling false positive-level (FPL) for siRNAs with no effects (A, B) and with extremely weak effects (C, D) for down-regulation in a primary screen without replicates. The legends in (B–D) are the same as in (A).

the sample size is greater than 16 (black curves). If the FPL with respect to $\beta_2 = 0.25$ is controlled to be 0.01 (Figure 5.3D), then the curves of FNLs with respect to $\beta_1 = -1.28, -1.645, -2, -3$ go down relatively quickly when the sample size is less than 24 and become relatively flat when the sample size is greater than 24. A sample size of 24 leads to an FNL of nearly 0.10 with respect to $\beta_1 = -3$ when controlling FPL $= 0.01$ with respect to $\beta_2 = -0.25$.

Figure 5.3 also reveals that, with an experimentally manageable FPR, the FNL for the siRNAs with extremely strong effects is very low, even if sample size is small (grey solid curves in Figure 5.3). The FNL for the siRNAs with very strong effects is also reasonably low as long as the sample size is greater than 16 (black solid curves in Figure 5.3). However, the FNL for siRNAs with strong, fairly strong, or moderate effects can be high. For example, even if 300 wells are occupied by a negative reference, the FNLs for siRNAs with strong, fairly strong, and moderate effects are approximately 0.35, 0.53, and 0.71, respectively (black dashed curves in Figure 5.3C), when controlling FPL = 0.025 for siRNAs with extremely weak effects. Therefore, the primary screen without replicates in 384-well plates does not have power large enough to detect siRNAs with strong, fairly strong, and moderate effects when controlling a manageable FPR for the siRNAs with extremely weak or no effects.

In a primary screen, we may also use the majority of the sample siRNAs as the negative reference group [174;176;180]. In such a case, n_- equals approximately 200 to 340. Compared with the case of $n_- = 20$, the FNRs of missing hits with very strong effects are reduced by about one half (solid black curves in Figure 5.3), but the amount of reduction in the FNRs of missing hits with strong, fairly strong, or moderate effects is not substantial (three types of dashed black curves in Figure 5.3). On the other hand, the use of the majority of sample siRNAs as the negative reference group is based on the assumption that the majority of sample siRNAs have no or tiny effects. This assumption may not be true in some screens.

In summary, a primary screen using 384-well plates should have at least 16 wells per plate, and an arrangement of 20 or 24 wells per plate is preferable for the negative control to be used as a negative reference for hit selection [175].

5.5 Applications

Here we demonstrate how to use various metrics including mean difference (or equivalently average fold change), percent inhibition, z-score, z^*-score, and SSMD to select hits in the HCV siRNA primary screen described in Section 1.4 of Chapter 1. In that screen, approximately 22,000 siRNA pools were tested across ninety-seven 384-well plates. Considering that the majority of siRNAs in the primary screen should have weak or no effect, we use the sample wells in a plate as the negative reference for calculating metrics for hit selection. Because the variability between plates is very large (Figure 5.4A), we adopt plate-wise analysis instead of experiment-wise analysis (Section 5.2.2).

From the raw data shown in Figure 5.4A, the measured values (in log2 scale) in the first nine plates are very different from the rest. The measured values of positive controls are slightly higher and the measured values of negative controls and sample wells are much lower in the first nine plates than in the remaining plates; this is clearly caused by a systematic shift of values in the sample wells and negative control wells. Consequently, the values of percent inhibition have a much larger spread in the first nine plates than in the remaining plates (Figure 5.4B). If we use the criterion of selecting siRNAs with percent inhibition above 40 as hits, we would select 215 hits, 92 (i.e., 43%) of which come from the first 9 plates (Figure 5.4B). If we did not check the raw data and directly used percent inhibition to select hits, we may have concluded that there were enriched true hits in the first nine plates, which is very misleading.

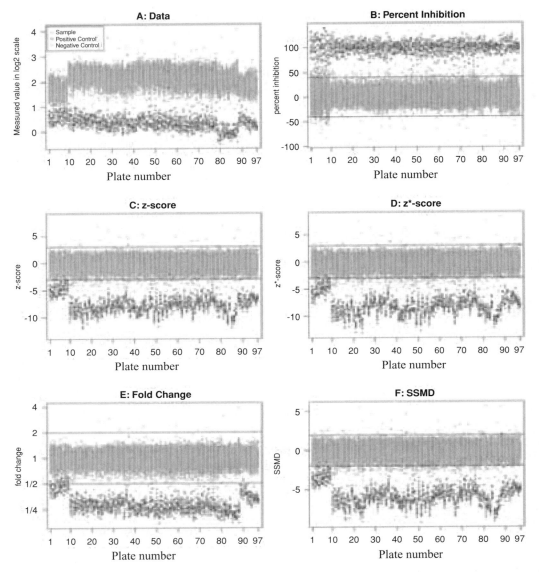

Figure 5.4   (See color insert following page 110.) Metrics for selecting hits in a hepatitis C virus RNAi primary screen without replicates.

z-score and z^*-score. The values of z-score and z^*-score are shown in Figure 5.4C and D, indicating that overall the z^*-score values are larger than the z-score values. If we use the criterion of selecting siRNAs with z-score less than −3 as inhibition hits, we would select 133 hits, as compared with 210 hits, by using the criterion of z^*-score less than −3. Among the 133 hits selected by z-score, 132 were also selected by z^*-score. Compared with siRNAs with weak or no effects, the true hits with large effects behave like outliers. Consequently, the estimated standard deviation will be inflated, and z-score will shrink. MAD is robust to the existence of outliers.

Therefore, the hit selection results using z^*-score are more reliable than those using z-score. So is the robust version of SSMD calculated using Formula E2 of Table 5.1.

SSMD. Figure 5.4E and F shows the difference (or equivalently fold change in log2 scale) and SSMD between an siRNA and mean of the negative reference with the mean calculated in the same way as for SSMD, as described in Section 5.2. That is, we calculated the mean and SD of the sample wells, excluding outliers in each plate. The R codes to calculate the difference and SSMD using sample wells as the negative reference are basically as follows:

```
for(i in 1:length(plates)) {
dataIn.df = data.df[plate.vec == plates[i],]
theInten.vec = dataIn.df[, Intensity]
theWells = dataIn.df[, wellName]
condtSample = theWells == sampleName
theIntenSample.vec = theInten.vec[condtSample]
theIntenSample.vec = theIntenSample.vec[!is.na(theIntenSample.vec)]
boxplot.stat = boxplot(theIntenSample.vec, plot = F)$stats[c(1,5)]
condt = theIntenSample.vec < boxplot.stat[2] &
    theIntenSample.vec > boxplot.stat[1]
negCenter = mean(theIntenSample.vec[condt])
negSpread = sd(theIntenSample.vec[condt])
theDiff.vec = theInten.vec — negCenter
zScore.vec = theDiff.vec/negSpread
ssmd.vec = z.vec/sqrt(2)
theResult.df = data.frame("log2Fold" = theDiff.vec,
"zScore" = zScore.vec, "ssmd" = ssmd.vec)
if(i == 1) {result.df = theResult.df}else {
    result.df = rbind(result.df, theResult.df)}
}
```

Fold change. When we use fold change to select hits, one big issue is the determination of a meaningful and practical cutoff for fold change. Conventionally, researchers have used the cutoff of two-fold change. However, that cutoff has no strong theoretical basis and is only applicable in some screens. In this example, using the cutoff of two-fold change in the inhibition direction, we only identified 21 inhibition hits as compared with at least 133 inhibition hits using other criteria. Another major issue is that fold change cannot take data variability into account, as illustrated in Figure 5.1B and C. This issue also exists in this screen. For example, because the measured values in plates 89 through 92 are compressed with smaller data variability (Figure 5.4A), the fold changes in those plates tend to be smaller than those in other plates; consequently, true hits in those plates would be missed if using fold changes to select hits.

Table 5.6. The size of siRNA effects classified using SSMD in the HCV primary screen

Effect Subtype	Activation		Inhibition	
	SSMD Cutoff	Count	SSMD Cutoff	Count
Extremely strong	$SSMD \geq 5$	3	$SSMD \leq -5$	2
Very strong	$5 > SSMD \geq 3$	9	$-5 < SSMD \leq -3$	29
Strong	$3 > SSMD \geq 2$	38	$-3 < SSMD \leq -2$	259
Fairly strong	$2 > SSMD \geq 1.645$	150	$-2 < SSMD \leq -1.645$	423
Moderate	$1.645 > SSMD \geq 1.28$	640	$-1.645 < SSMD \leq -1.28$	957
Fairly moderate	$1.28 > SSMD \geq 1$	1,280	$-1.28 < SSMD \leq -1$	1,285
Fairly weak	$1 > SSMD \geq 0.75$	2,081	$-1 < SSMD \leq -0.75$	1,808
Weak	$0.75 > SSMD > 0.5$	3,161	$-0.75 < SSMD < -0.5$	2,497
Very weak	$0.5 \geq SSMD > 0.25$	3,883	$-0.5 \leq SSMD < -0.25$	3,250
Extremely weak	$0.25 \geq SSMD > 0$	4,111	$-0.25 \leq SSMD < 0$	3,716
Zero	$SSMD = 0$	0		

Note: "Count" is the number of siRNAs in each category.

The values of SSMD are robust to the shift and compression of data, as shown in Figure 5.4F. In addition, a unique advantage to using SSMD is that we can use it to classify the size of siRNA effects. If the true value equals the estimated value of SSMD, then we may use the criteria in Table 5.2 and obtain the classifying results in Table 5.6 for siRNA effects in the HCV RNAi primary screen. Based on the results in Table 5.6, we may focus on the 713 siRNAs with fairly strong or stronger inhibition effects and the 200 siRNAs with fairly strong or stronger activation effects in the follow-up studies.

The SSMD-based classifying criterion works effectively when the sample size is large. However, in the primary screen without replicates, there is only one replicate for an siRNA, although there are approximately 300 replicates for the negative reference. Because the sample size is not large and the true values of SSMD are unknown, we need to determine a cutoff of estimated SSMD values for hit selection to control FPRs and FNRs at the desired level. As presented in Section 5.4, based on SSMD, we can control FPR not only with respect to no effects, but also with respect to extremely weak effects. The FPRs with respect to true SSMD values of 0 and −0.25 and the FNRs with respect to true SSMD values of −2, −3, and −5 are shown in Figure 5.5.

The HCV screen [174;180] of Figure 5.4 had one negative control and two positive controls, a stronger inhibition control, and a weaker inhibition control. The main interest is in the inhibition direction. Figure 5.5 shows the error rates for selecting inhibition hits. We use the negative control to calculate empirical FPR (black solid line in Figure 5.5) and the two inhibition controls to calculate corresponding empirical FNRs (grey and black dashed lines in Figure 5.5). The empirical FPR in a decision rule for hit selection based on a negative control is the ratio of the number of wells

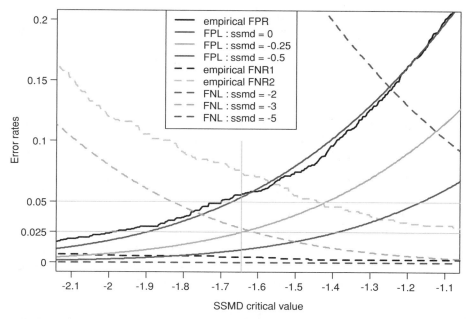

Figure 5.5   (See color insert following page 110.) False-positive levels (FPLs), false-negative levels (FNLs),
and empirical false-negative rates (FNRs) for selecting inhibition hits in a hepatitis C virus RNAi
primary screen without replicates.

being selected as hits to the total number of wells for this negative control. The
empirical FNR in a decision rule for hit selection based on a positive control is the
ratio of the number of wells being selected as a non-hit to the total number of wells
for this positive control. The empirical FPR curve for the negative control is nearly
the same as the theoretical FPR curve with respect to SSMD = −0.5 (blue solid line
in Figure 5.5), which indicates that the negative control actually has weak inhibition
effects. The empirical FNR curve for the weaker inhibition control (dashed grey line)
lies between the theoretical FNR curves with respect to SSMD = −2 and SSMD =
−3, respectively (green and blue dashed lines in Figure 5.5), which indicates that the
weaker inhibition control has a strong effect, with a theoretical SSMD value between
−2 and −3. Similarly, we can infer that the stronger inhibition control has a very
strong inhibition effect, with a theoretical SSMD value between −3 and −5.

In this screen, to control FPR with respect to SSMD = −0.25 to be less than 0.025
in the inhibition direction, the SSMD cutoff for selecting inhibition hits should be
no greater than −1.645. The use of an SSMD cutoff of −1.645 leads to 713 selected
inhibition hits. The theoretical error rates corresponding to this cutoff are less than
0.01 for FPR with respect to SSMD = 0, less than 0.025 for FPR with respect to SSMD
= −0.25, less than 0.054 for FPR with respect to SSMD = −0.5, less than 0.308 for
FNR with respect to SSMD = −2, less than 0.028 for FNR with respect to SSMD =
−3, and less than 0.000001 for FNR with respect to SSMD = −5. The SSMD cutoff
of −1.645 leads to an empirical FNR of 0.076 for the weaker inhibition control

and 0.0045 for the stronger inhibition control and an empirical FPR of 0.0567 for the negative control. Accordingly, we can obtain the FNRs and FPRs for the 200 activation hits obtained using the SSMD cutoff of 1.645 in the activation direction.

5.6 Conclusions

In many RNAi screens, hit selection is the ultimate goal and final stage in data analysis before biological pathway analysis. As shown in this chapter and then again in Chapter 6, there are many methods for hit selection, and different methods may produce different results. We need to adopt suitable analytic methods for hit selection. Because SSMD is effective in measuring the size of siRNA effects, capable of capturing data variability and robust to outliers, I advocate the use of SSMD for hit selection in RNAi screens. In addition, the reader should keep in mind that the results of hit selection are affected not only by the methods of hit selection described in this chapter, but also by experimental design, data normalization, and quality control, as described in Chapters 2, 3, and 4. Therefore, we should follow the strategies and adopt suitable methods in experimental design, data normalization, and quality control as described in Chapters 2, 3, and 4 before we conduct formal hit selection using the methods described in this chapter and in the next chapter.

6

Hit Selection in Genome-Scale RNAi Screens with Replicates

In Chapter 5, we discussed analytic methods for hit selection in screens without replicates. Analytic methods for hit selection in screens with replicates differ from those without replicates, mainly because we can directly estimate data variability for a tested siRNA based on multiple measured values of a phenotype for an siRNA in a screen with replicates, but we cannot do so for a screen without replicates. In a primary screen without replicates, we must make a strong assumption that each siRNA has the same variability as a negative reference group in a plate and use the variability of this negative reference to represent the variability of each siRNA. In a screen with replicates, the analytic methods do not rely on this assumption, and thus we can use more powerful methods.

In this chapter, I present analytic methods for hit selection in screens with replicates. Specifically, I provide metrics for hit selection in screens with replicates in Section 6.1, in which the focus is on the classical t-statistic and the SSMD method. In Section 6.2, I present a dual-flashlight plot in which both mean difference and SSMD are displayed, and in Section 6.3, I elaborate on various decision rules and associated false-positive and false-negative rates. In Section 6.4, I explore false discovery and false nondiscovery rates; in Section 6.5, I investigate sample size determination in screens with replicates; and in Section 6.6, I present SSMD-based statistical methods for adjusting for off-target effects. Finally, I demonstrate how to use the analytic methods in real examples in Section 6.7, with a general discussion in Section 6.8.

6.1 Metrics for Hit Selection in Screens with Replicates

In a screen with replicates, a potential approach for selecting hits is to adopt methods similar to those for primary screens without replicates, as follows:

(i) Choose a metric for a primary screen without replicates.
(ii) Calculate the value of the chosen metric based on its corresponding formula from A1 through E1 or A2 through E2 in Table 5.1 of Chapter 5 for each replicate of an siRNA.

(iii) Use their mean or median across replicates for this siRNA as a metric for selecting hits.

One major issue with this approach is that it does not take into account data variability of an siRNA across its replicates, although the z-score in D1 or D2 of Table 5.1 accounts for data variation in different plates. Therefore, the better approach is to adopt statistical methods, such as the classical t-statistic or SSMD, that account for variability across replicates of an siRNA, which is described in this section.

6.1.1 *t*-Statistic

In primary or confirmatory HTS experiments with replicates, a t-test for testing mean difference has been used for selecting hits. It is well known that when there are replicates, a t-test is better than the z-score method for testing no mean difference in two groups, especially when the sample size is small. Because plate-to-plate variability is usually higher than within-plate variability, a paired t-test is often used for hit selection. That is, for the ith siRNA with n replicates, we calculate the difference between the measured value (usually in log scale) of the siRNA and the mean or median value of a negative control in a plate, d_{ij} ($j = 1, \ldots, n$), then calculate the corresponding t-statistic value as follows:

$$t\text{-statistic} = \frac{\bar{d}_i}{\frac{s_i}{\sqrt{n}}} \qquad (6.1)$$

where $\bar{d}_i$ and s_i are, respectively, the sample mean and standard deviation of d_{ij} for the ith siRNA. We also calculate its corresponding p-value based on the t-distribution, namely, the above t-statistic has a central t-distribution with n-1 degrees of freedom under the null hypothesis of zero mean difference.

One issue with the use of the t-value and p-value is that they are affected by both sample size and siRNA effects. People tend to think that a small p-value indicates a large siRNA effect and a large p-value indicates a small or no siRNA effect. However, as demonstrated in Figure 6.1, this is not true. An siRNA with a large p-value may have a large effect (Figure 6.1A1). On the other hand, an siRNA with a small p-value may have a small effect (Figure 6.1B2).

6.1.2 SSMD Method

In RNAi screens, we are really most interested in the size of siRNA effects. Thus we need to separate effect size from the impact of sample size. The t-statistic and associated p-value come from testing no mean difference and are affected by both sample size and effect size. They are not designed to measure the size of siRNA effects. SSMD directly assesses the size of siRNA effects [161;162;174]. In a screen with n replicates, a simple estimate of SSMD is:

$$\text{SSMD} = \frac{\bar{d}_i}{s_i} \qquad (6.2)$$

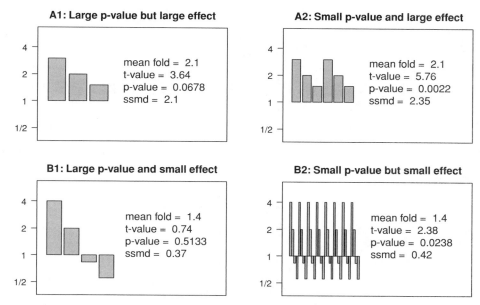

Figure 6.1   Examples of data for an siRNA, demonstrating the use of fold change, *t*-value, *p*-value, and SSMD in RNAi screens with replicates. One bar represents one replicate of the difference between an siRNA and a negative reference in a plate. The data in (A2) come from repeating the data in (A1) two times, and the data in (B2) come from repeating the data in (B1) eight times.

where $\bar{d}_i$ and s_i are, respectively, the sample mean and standard deviation of the difference between the ith siRNA and a negative reference. Another estimate, called uniformly minimal variance unbiased estimate of SSMD, is:

$$\text{SSMD} = \frac{\Gamma\left(\dfrac{n-1}{2}\right)}{\Gamma\left(\dfrac{n-2}{2}\right)}\sqrt{\frac{2}{n-1}}\frac{\bar{d}_i}{s_i} \qquad (6.3)$$

where $\Gamma\left(\cdot\right)$ is a gamma function. See Chapter 8 for more details about the mathematical formulation of these SSMD estimates in screens with replicates.

As demonstrated in Figure 6.1, the values of SSMD correctly indicate that the siRNA in A1 and A2 has a strong effect and the siRNA in B1 and B2 has a very weak effect. An additional example is shown in siRNA B1 and siRNA B2 in Figure 6.2, in which mean fold change is large, but effect size is small. SSMD correctly indicates that the two siRNAs have weak effects. In addition, one major advantage of using SSMD is that there is a theoretical basis for deriving thresholds of SSMD population values to assess the strength of siRNA effects, as shown in Table 5.2 of Chapter 5 [167].

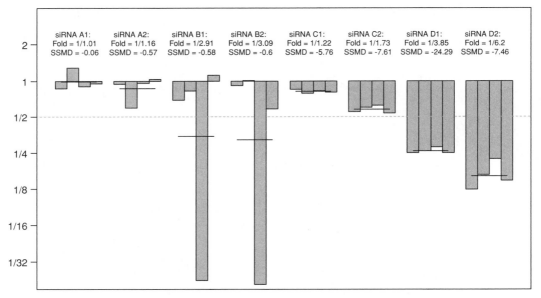

Figure 6.2   Examples of data for eight siRNAs each with four replicates in a confirmatory screen. siRNAs A1 and A2 have a small mean fold change and a small absolute value of SSMD, siRNAs B1 and B2 have a large mean fold change and a small absolute value of SSMD, siRNAs C1 and C2 have a relatively small mean fold change and a large absolute value of SSMD, and siRNAs D1 and D2 have a large mean fold change and a large absolute value of SSMD.

6.2 Dual-Flashlight Plot

SSMD can overcome the drawback of fold change not being able to capture data variability. On the other hand, because SSMD is the ratio of mean to standard deviation of a variable, we may get a large SSMD value when the standard deviation is very small, even if the mean is small (siRNA C1 in Figure 6.2). In some cases, a too small mean value may not have a biological impact. As such, siRNAs with large SSMD values (or differentiations) but too small mean values may not be of interest. The issue can be more serious if the too large estimated values of SSMD are caused by too small sample variance due to a small sample size. Too small sample variance causes a similar issue in the use of p-values for measuring differential expression. To address this issue, a plot called a *dual-flashlight plot* has been proposed to display both SSMD and mean of difference, similar to a volcano plot [170].

Figure 6.3 is a dual-flashlight plot showing both estimated means and SSMDs of the variable for the difference between an siRNA and a negative control in a simulated experiment of 25,000 siRNAs, each with 50 replicates. As a whole, the points in a dual-flashlight plot look like the beams of a flashlight with two heads. With the dual-flashlight plot, we can see how the siRNAs are distributed into each category of differentiation, as shown in Figure 6.3. Meanwhile, we can exclude the siRNAs with large estimated SSMD values but too small fold change values. For example, if a mean fold change between 1/1.4 and 1.4 does not have significant biological impact,

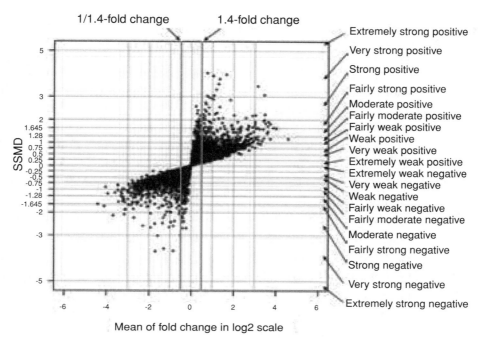

Figure 6.3   Dual-flashlight plot of a variable representing the difference between an siRNA and a negative control in a simulated experiment of 25,000 siRNAs, each with 50 replicates. The gray
horizontal lines represent the thresholds in the SSMD-based classifying criteria.

we may exclude all the genes between the two thick gray vertical lines of Figure 6.3,
even if some of them have an estimated SSMD value greater than 2 or less than −2.

A clear benefit of the dual-flashlight method is demonstrated in Figure 6.2. From
the bar plots in Figure 6.2, siRNAs A1, A2, B1, and B2 do not have consistently large
effects of interest; siRNA C1 has consistently weak effects, which may be of interest
in some cases and may not be in other cases; and siRNAs C2, D1, and D2 clearly
have consistently large effects of interest. Using the dual-flashlight method, we can
readily select siRNAs A1, A2, B1, B2, and C1 as non-hits and select siRNAs C2, D1,
and D2 as hits.

The SSMD-based classifying criteria shown in Figure 6.3 (which are also displayed
in Table 5.2 of Chapter 5) are based on the true value (i.e., population value in a
distributional level) of SSMD for each siRNA. However, the true value of SSMD is
usually unknown for an siRNA. Thus these criteria work effectively when the sample
size is large and work only approximately when the sample size is small. When the
sample size is small, a more accurate method is to find an SSMD-based decision
rule to achieve a desired FPR for controlling the number of siRNAs with weak or no
effects in the selected hits and a desired FNR for controlling the number of siRNAs
with large effects not being selected as hits. In the following section, I will first
describe classical decision rules based on t-statistic and then present decision rules
based on SSMD.

6.3 Decision Rules for Hit Selection in Screens with Replicates

The true value of SSMD can effectively assess the strength of siRNA effects and can classify them into different categories, such as zero, extremely weak, very weak, and so on (as shown in Table 5.2 of Chapter 5). Consequently, based on estimated values of SSMD, we can control the FPR not only with respect to no effect, but also with respect to extremely weak effects or even very weak effects. Similarly, we can control the FNR with respect to extremely strong, very strong, and strong effects. Using the SSMD-based definition of false positives and false negatives as described in Section 5.3.1 of Chapter 5, we can explore the FPR and FNR in SSMD-based decision rules for hit selection in RNAi screens with replicates.

In screens with replicates, the SSMD-based decision rule for selecting down-regulated hits is as follows.

Decision rule 6.1 (down-regulation, SSMD-based). Declare an siRNA as a hit if it has an estimated SSMD value less than or equal to a critical value β^* and as a non-hit otherwise.

The formulas for calculating FPRs and FNRs in decision rule 6.1 are shown in the left panel of Table 6.1.

For an individual siRNA, if we treat its estimated value of SSMD β_{obs} as a critical value in a decision rule, the corresponding FPL becomes the p-value of this siRNA. Note that the FPL is the maximal FPR in a decision rule. In decision rule 6.1, FPR $= \Pr(\hat{\beta} \leq \beta^* | \beta \geq \beta_2)$. Given $\beta \geq \beta_2$, the maximal FPR is the FPL, which is achieved at $\beta = \beta_2$, namely, FPL $= \Pr(\hat{\beta} \leq \beta^* | \beta = \beta_2)$. The p-value for an individual siRNA with estimated SSMD β_{obs} is p-value $= \Pr(\hat{\beta} \leq \beta_{\mathrm{obs}} | \beta = \beta_2)$. Similar to the definition of the p-value based on FPL, we may define the p^*-value for an individual siRNA based on the FNL when we treat the estimated value of SSMD of this siRNA as a critical value in a decision rule. That is, FNL $= \Pr(\hat{\beta} \geq \beta^* | \beta = \beta_1)$ in decision rule 6.1 and p^*-value $= \Pr(\hat{\beta} \geq \beta_{\mathrm{obs}} | \beta = \beta_1)$ for an individual siRNA with estimated SSMD β_{obs}. The formulas for calculating p-value and p^*-value for an siRNA are also listed in the left panel of Table 6.1.

Similarly, the SSMD-based decision rule for selecting up-regulated hits in screens with replicates is as follows.

Decision rule 6.2 (up-regulation, SSMD-based). Declare an siRNA as a hit if it has an estimated SSMD value greater than or equal to a critical value β^* and as a non-hit otherwise.

The formulas for calculating FPRs, FNRs, p-values, and p^*-values in decision rule 6.2 are also shown in the left panel of Table 6.1. More mathematical details can be found in Chapter 8.

Traditionally, the classical t-test is used for hit selection in RNAi screens with replicates. The classical t-test uses mean difference to define true hits, false positives, and false negatives in the same manner as those in the z-score method as described

Table 6.1. Decision rule, FPL, FNL, and p-value for hit selection in RNAi primary screens with replicates

	Calculation Formula	
Direction	SSMD Method	Classical t-Test
Down-regulation	**Decision Rule 6.1**: Any siRNA is a hit if estimated SSMD $\leq \beta^*$ and a non-hit otherwise $$\text{FPL} = F_{t(n-1,\sqrt{n}\beta_2)}\left(\frac{\beta^*}{k}\right)$$ $$\text{FNL} = 1 - F_{t(n-1,\sqrt{n}\beta_1)}\left(\frac{\beta^*}{k}\right)$$ $$p\text{-value} = F_{t(n-1,\sqrt{n}\beta_2)}\left(\frac{\beta_{\text{obs}}}{k}\right)$$ $$p^*\text{-value} = 1 - F_{t(n-1,\sqrt{n}\beta_1)}\left(\frac{\beta_{\text{obs}}}{k}\right)$$	**Decision Rule 6.3**: Any siRNA is a hit if it has t-value $\leq t^*$ and a non-hit otherwise $$\text{FPL} = F_{t(n-1)}\left(t^*\right)$$ $$\text{FNL} = 1 - F_{t(n-1)}\left(t^*\right)$$ $$p\text{-value} = F_{t(n-1)}\left(t_{\text{obs}}\right)$$
Up-regulation	**Decision Rule 6.2**: Any siRNA is a hit if estimated SSMD $\geq \beta^*$ and a non-hit otherwise $$\text{FPL} = 1 - F_{t(n-1,\sqrt{n}\beta_2)}\left(\frac{\beta^*}{k}\right)$$ $$\text{FNL} = F_{t(n-1,\sqrt{n}\beta_1)}\left(\frac{\beta^*}{k}\right)$$ $$p\text{-value} = 1 - F_{t(n-1,\sqrt{n}\beta_2)}\left(\frac{\beta_{\text{obs}}}{k}\right)$$ $$p^*\text{-value} = F_{t(n-1,\sqrt{n}\beta_1)}\left(\frac{\beta_{\text{obs}}}{k}\right)$$	**Decision Rule 6.4**: Any siRNA is a hit if it has t-value $\geq t^*$ and a non-hit otherwise $$\text{FPL} = 1 - F_{t(n-1)}\left(t^*\right)$$ $$\text{FNL} = F_{t(n-1)}\left(t^*\right)$$ $$p\text{-value} = 1 - F_{t(n-1)}\left(t_{\text{obs}}\right)$$

Note: FNL, false-negative level; FPL, false-positive level. $F_{t(n-1)}(\cdot)$ and $F_{t(n-1,\sqrt{n}\beta)}(\cdot)$ are the cumulative distribution functions of central t-distribution $t(n-1)$ and noncentral t-distribution $t(n-1, \sqrt{n}\beta)$, respectively; n is the number of replicates; $k = \sqrt{1/n}$ when SSMD is estimated using Formula 6.2, and $k = \Gamma(\frac{n-1}{2})/\Gamma(\frac{n-2}{2})\sqrt{\frac{2}{n(n-1)}}$ when SSMD is estimated using Formula 6.3. t_{obs} and β_{obs} are t-value and estimated SSMD value of an siRNA, respectively. β_1 and β_2 are population values of SSMD to indicate large and small effects, respectively. See Chapter 8 for how all these formulas are derived.

in Section 5.3.1 of Chapter 5. The decision rule based on the t-test for selecting down-regulated hits in screens with replicates is as follows.

Decision rule 6.3 (down-regulation, t-test). Declare an siRNA as a hit if it has a value t of t-statistic for the same sample size (as calculated in Formula 6.1) less than a critical value t^* and as a non-hit otherwise.

The decision rule based on t-test for selecting up-regulated hits in screens with replicates is as follows.

Decision rule 6.4 (up-regulation, t-test). Declare an siRNA as a hit if it has a value t of t-statistic for the same sample size (as calculated in Formula 6.1) greater than a critical value t^* and as a non-hit otherwise.

Table 6.2. SSMD-based decision rules, FPL, and FNL in RNAi screens with replicates when a preset FPL or FNL is fixed

	Calculation Formula	
Direction	Control FPL w.r.t. β_2 be α_1	Control FNL w.r.t. β_1 to be α_2
Down-regulation	**Decision Rule 6.5**: Any siRNA is a hit if estimated SSMD $\leq \beta_{\alpha_1}$ and a non-hit otherwise, where β_{α_1} is obtained by solving $$F_{t(n-1,\sqrt{n}\beta_2)}\left(\frac{\beta_{\alpha_1}}{k}\right) = \alpha_1$$ **Formula A:** $$FNL = 1 - F_{t(n-1,\sqrt{n}\beta_1)}\left(\frac{\beta_{\alpha_1}}{k}\right)$$	**Decision Rule 6.6**: Any siRNA is a hit if estimated SSMD $\leq \beta_{\alpha_2}$ and a non-hit otherwise, where β_{α_2} is obtained by solving $$F_{t(n-1,\sqrt{n}\beta_1)}\left(\frac{\beta_{\alpha_2}}{k}\right) = 1 - \alpha_2$$ **Formula B:** $$FPL = F_{t(n-1,\sqrt{n}\beta_2)}\left(\frac{\beta_{\alpha_2}}{k}\right)$$
Up-regulation	**Decision Rule 6.7**: Any siRNA is a hit if estimated SSMD $\geq \beta_{\alpha_1}$ and a non-hit otherwise, where β_{α_1} is obtained by solving $$F_{t(n-1,\sqrt{n}\beta_2)}\left(\frac{\beta_{\alpha_1}}{k}\right) = 1 - \alpha_1$$ **Formula C:** $$FNL = F_{t(n-1,\sqrt{n}\beta_1)}\left(\frac{\beta_{\alpha_1}}{k}\right)$$	**Decision Rule 6.8**: Any siRNA is a hit if estimated SSMD $\geq \beta_{\alpha_2}$ and a non-hit otherwise, where β_{α_2} is obtained by solving $$F_{t(n-1,\sqrt{n}\beta_1)}\left(\frac{\beta_{\alpha_2}}{k}\right) = \alpha_2$$ **Formula D:** $$FPL = 1 - F_{t(n-1,\sqrt{n}\beta_2)}\left(\frac{\beta_{\alpha_2}}{k}\right)$$

Note: FNL, false-negative level; FPL, false-positive level; w.r.t., with respect to. $F_{t(n-1,\sqrt{n}\beta)}(\cdot)$ is the cumulative distribution function of noncentral t-distribution $t(n - 1, \sqrt{n}\beta)$; n is the number of replicates; $k = \sqrt{1/n}$ when SSMD is estimated using Formula 6.2, and $k = \Gamma(\frac{n-1}{2})/\Gamma(\frac{n-2}{2})\sqrt{\frac{2}{n(n-1)}}$ when SSMD is estimated using Formula 6.3. β_1 and β_2 are true values of SSMD to indicate large and small effects, respectively.

The error rates in decision rules 6.3 and 6.4 are shown in the right panel of Table 6.1.

In RNAi screens, it is usually desirable to control the FPR to be less than a level α_1 where α_1 equals 0.05 or 0.01. Based on the formulas in Table 6.1, to control the FPL with respect to β_2 to be α_1, we need to set $\beta^* = \beta_{\alpha_1}$ where β_{α_1} is a critical value such that $F_{t(n-1,\sqrt{n}\beta_2)}(\beta_{\alpha_1}/k) = \alpha_1$. Similarly, we control the FNR with respect to β_1 to be less than a pre-set level α_2 and calculate the corresponding critical value and the FPL. The formulas are shown in Table 6.2.

6.4 False Discovery Rate, False Non-Discovery Rate, *q*-Value, and *q**-Value

Whether using mean difference or SSMD, the methods for adopting FPR, FNR, *p*-value, and *p**-value to control false positives and false negatives are based on a single test. Given that a large number of siRNAs are tested in a genome-scale RNAi screen, the FPR will be inflated. Hence one issue in these methods is the adjustment of error rates in multiple hypothesis testing. The simplest adjustment may be the Bonferroni correction [15], which is the FPR or *p*-value for a single test divided by the total number of tests conducted. Currently, the most effective method for adjusting for the multiplicity issue is the use of the false discovery rate (FDR) [11;144] and false non-discovery rate (FNDR) [52].

More importantly, as pointed out by Storey and Tibshirani [146], FPR and FDR are often mistakenly equated, but their difference is very important. Given a particular selection criterion or a decision rule for selecting siRNAs with large effects, FDR is the proportion of the selected hits (i.e., discoveries) that are true non-hits, whereas FPR is the proportion of all the true non-hits in the study selected as hits. FNDR is the proportion of all the declared non-hits (i.e., non-discoveries) that are true hits, whereas FNR is the proportion of all the true hits in the study declared as non-hits. For example, an FPR of 5% means that, on average, 5% of all siRNAs with no or small effects in the study will be selected as hits. By contrast, an FDR of 5% means that among the list of selected hits, on average, 5% are siRNAs with no or small effects. An FNR of 10% means that, on average, 10% of all siRNAs with large effects in the study will be declared as non-hits. By contrast, an FNDR of 10% means that among the list of declared non-hits, on average, 10% are siRNAs with large effects. FPR and FNR are based on unknown numbers of true hits and true non-hits in a study, whereas FDR and FNDR are based on known numbers of selected hits and declared non-hits in a study. Therefore, FDR and FNDR provide more useful information in a study than FPR and FNR.

Consider the problem of simultaneously testing m siRNAs, of which m_0 are true non-hits. False positives are the true non-hits among the selected hits, and false negatives are true hits among the declared non-hits. FPR equals the expectation of the total number of false positives FP divided by the total number of tests m, namely $E(FP/m)$. By contrast, FDR equals the expectation of FP divided by the total number of selected hits R, namely $E(FP/R)$. Similarly, FNR equals the expectation of the total number of false negatives FN divided by the total number of tests m, namely $E(FN/m)$. FNDR is the expectation of FN divided by the total number of declared non-hits $m - R$, namely $E(FN/(m - R))$.

The well-known q-value is a term defined similarly to p-value. Whereas a p-value is defined on FPR, a q-value is equivalently defined on FDR. When using SSMD for selecting up-regulated hits, for a particular siRNA with an observed SSMD value β_{obs}, the p-value equals the maximum probability of selecting this siRNA as a hit (given that the true value of SSMD is less than or equal to a value β_2) when we use the following selection criterion: any siRNA is selected as a hit if it has an estimated SSMD value greater than or equal to β_{obs} and as a non-hit otherwise. That is, the p-value equals the maximal FPR when we use decision rule 6.1 with β_{obs} as a critical value. Equivalently, q-value for this siRNA with respect to β_2 is the maximum FDR using β_{obs} as a critical value. Similarly, the p^*-value is the maximal FNR with respect to β_1, and equivalently, the q^*-value for this siRNA with respect to β_1 is the maximum FNDR using β_{obs} as a critical value.

There are an impressive number of algorithms for estimating the q-value and/or controlling FDR in the literature [3;22;35;42;57;90;93;103;120;124;144;146]. One

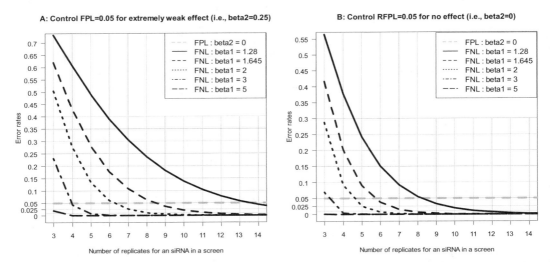

Figure 6.4   The changes of error rates versus sample size by controlling FPL for siRNAs with extremely weak effects (A) and with no effects (B) in a confirmatory screen.

popular algorithm is the Benjamin-Hochberg procedure [11]. The FDR calculated using the Benjamin-Hochberg procedure is conservative [146;147]. After obtaining the p-value using the formulas listed in Table 6.1, we can use existing R packages (e.g., *qvalue* [146], *multtest* [11] or *fdrtool* [147]) to calculate the q-value with respect to $\beta \geq \beta_2$ in the down-regulated direction when using SSMD. To calculate the q^*-value in the down-regulated direction, we can treat the p^*-value (using the formulas listed in Table 6.1) as the p-value for testing null hypothesis $H_0 : \beta \leq \beta_1$ and calculate the corresponding q-value; the resulting q-value equals the q^*-value with respect to $\beta \leq \beta_1$. Similarly, we can derive FDR, FNDR, q-value, and q^*-value for hit selection in the up-regulated direction. Additional mathematical details about FDR, FNDR, q-value, and q^*-value and their estimation is provided in Section 8.6 of Chapter 8.

6.5 Sample Size Determination

In a confirmatory screen, the goal is to achieve a reasonably low FNR of missing siRNAs with large effects (i.e., to obtain a reasonably high power of selecting siRNAs with large effects) while maintaining a pre-set low level (such as 0.05) of the FPR of selecting siRNAs with extremely weak or no effects. Figure 6.4 displays the FNLs with respect to SSMD critical values of 1.28, 1.645, 2, 3, and 5, respectively, with FPL = 0.05 with respect to $\beta_2 = 0.25$ (A) or $\beta_2 = 0$ (B). When FPL = 0.05 with respect to $\beta_2 = 0.25$, the FNL with respect to $\beta_1 = 5$ is 0.019, the FNL with respect to $\beta_1 = 3$ is 0.231, and the FNL with respect to $\beta_1 = 2$ is 0.505 (or 0.288) for a sample size of 3. Clearly, a sample size of three replicates can achieve a reasonably high power for detecting siRNAs with extremely strong effects (bottom black dashed curve of Figure 6.4) and may achieve an acceptable power for detecting siRNAs with very strong effects (next-to-bottom black dashed curve), but cannot achieve an acceptable power for detecting siRNAs with strong, fairly strong, or moderate effects (top three black curves). These curves show that 4, 5, or 6 is a reasonable sample size for detecting siRNAs with strong effects; 5, 6, 7, or 8 is a reasonable sample size for detecting siRNAs with fairly strong effects (top black dashed curve); and 8, 9, 10, or 11 samples are best for detecting siRNAs with moderate effects (black solid curve of Figure 6.4).

Therefore, in a confirmatory screen, a sample size of at least four (i.e., the arrangement of at least four replicate plates per source plate) is required for detecting siRNAs with strong, fairly strong, or moderate effects. Regarding the tradeoff between benefit and cost, any sample size between 4 and 11 is a reasonable choice for selecting siRNAs with strong, fairly strong, or moderate effects. If the main focus is the selection of siRNAs with strong effects, a sample size of four or five is a good choice. If cost is not a serious consideration, then a sample size of six, seven, or eight is preferred, especially when only one or two sets of source plates are investigated in a confirmatory screen. If we want to have enough power to detect siRNAs with moderate effects, then the sample size needs to be 8, 9, 10, or 11 [175].

6.6 Analytic Methods Adjusting for Off-Target Effects

6.6.1 Introduction to Experiments Addressing Off-Target Effects

RNAi HTS is broadly used in the identification of genes associated with specific biological phenotypes [7;19;44;88;113;118;183]. The impact of an siRNA on a measured phenotype may come from two major sources. One is the knockdown of the targeted gene that plays a role in generating the phenotype, and the other is the knockdown of one or more unintended genes that affect the phenotype. The first is an on-target effect, and the second is an off-target effect. Off-target gene knockdown is an RNAi-mediated event in which unintended mRNA targets with sequence homology to the RNAi oligonucleotide are degraded. False positives generated by off-targets during phenotypic screens can lead to false leads and the use of resources to explore nonproductive research paths and may impede analysis of RNAi screens [78;87].

As described in Chapter 2, there are many approaches and designs for reducing the impact of off-target effects. However, as Echeverri [40] point out, although the ideal approach to saving experiments from off-target effects remains technically challenging in most contexts, a more accessible solution, adopted by most researchers and companies today, is to test multiple siRNAs with different sequences against a target gene to increase the level of confidence in positive hits.

With the consideration of controlling experimental cost, most genome-scale RNAi screening projects start with a primary screen where thousands of siRNA pools (three to four duplexes per well) are investigated without replicates. The hits (normally in hundreds) from the siRNA pools in the primary screen are further investigated in one or more confirmatory screens where each pool has replicates (normally 3 to 6 replicates). The genes selected in the confirmatory screen (normally under 100) from the confirmatory screens are further investigated by designing siRNA singles per gene in a screen. The screen in which the phenotypic effects of multiple siRNA singles per targeted gene are measured separately is called a deconvolution screen.

6.6.2 Issues in Current Analytic Methods Addressing Off-Target Effects

To capture the information of multiple siRNAs against a gene in an RNAi screen, the straightforward analytic method is the so called "frequency approach": first select hits based on the individual activity of each siRNA and then select genes based on the frequency with which the multiple siRNAs targeting this gene are selected as hits. One criterion that some people adopt in the frequency approach is as follows: a gene is selected as a hit if 25% of the siRNAs are selected as hits (e.g., two of seven siRNAs). The frequency approach has two major issues: (i) captures the information of only a portion of siRNA singles with strong effects and (ii) misses the selection of genes with consistent moderate effects.

Classical t-test is another option which tests whether the mean of all siRNAs targeting this gene being zero. The issue with the p-value from the classical *t*-test is that it is strongly affected by sample size: for the same size of non-zero effects, the large the sample size, the smaller the p-value. Consequently, the genes with fewer siRNAs are less likely to be selected as hits even though they may have a large effect.

More recently developed analytic methods include Konig et al.'s redundant siRNA activity (RSA) method [87] and Barbie et al.'s RNAi gene enrichment ranking (RIGER) method [6]. Both the RIGER and RSA methods examine the rank distribution of all siRNAs targeting a gene and calculate the statistical significance of all siRNAs targeting a gene being unusually distributed toward the top or bottom ranking slots. To derive the statistical significance, the RSA method uses an iterative hypergeometric distribution formula whereas the RIGER method uses a two-sample weighted likelihood ratio statistic combined with permutation. A feature of these ranking-based methods is that a gene with multiple moderately active siRNAs is weighted more heavily than a gene with fewer active siRNAs. The RSA method requires two arbitrary thresholds to initially define active siRNAs and negative siR-NAs whereas the RIGER method does not.

In both the RSA and RIGER methods, the null distribution is formalized from all the siRNAs in the entire list in an experiment. This null distribution is reasonable if most investigated siRNAs have no or very small effects. Otherwise, especially when most investigated siRNAs are active in one direction, this null distribution is problematic. Therefore, the RSA and RIGER methods are applicable to a primary screen where most siRNA singles should have no or very small effects; however, they are inapplicable to a deconvolution screen where most siRNAs are pre-selected to have up- or down-regulated effect.

6.6.3 SSMD Methods Assessing Collective Activity of Multiple siRNAs

Here I present an SSMD-based method [168] to capture the collective activity of multiple siRNAs per gene. This method that does not draw the null distribution from all investigated siRNAs in a screen; thus it works effectively for both primary and deconvolution screens with multiple siRNA singles against a gene. Like RSA and RIGER, the new method captures the collective activity of multiple singles against a gene, thus minimizing the impact of off-target effects. And a gene with multiple moderately active singles is weighted more heavily than a gene with fewer active singles in the new method. Therefore, the proposed method will not have the issues of the frequency approach. Like the t-statistic, the SSMD estimate captures both sample mean and sample variability of multiple investigated siRNAs targeting a gene. Unlike the t-statistic, the SSMD estimate is robust to sample size. The SSMD method naturally incorporates all the information of multiple siRNAs targeting the same gene in a strong statistical basis. In addition, the SSMD method can assess not only the collective activities of multiple siRNAs targeting the same gene, but also the strength of specific effect of each siRNA beyond its collective activity.

Different siRNAs targeting the same gene may have different potency and different silencing kinetics, leading to different specific on-target effects beyond shared on-target effects. Thus the on-target phenotypic effect of an siRNA can be partitioned into two parts: (i) a specific on-target effect that is unique to the siRNA, and (ii) a shared on-target effect that is common to siRNAs targeting the same gene.

One major reason for examining the collective activity of multiple siRNAs is that the off-target effects of these siRNAs are very likely to have different directions and thus may be canceled out in their collective activity, whereas the on-target effects of these siRNAs should be in the same direction and thus should not be canceled out with each other. Both average fold change (in log scale) and collective SSMD for multiple siRNAs are robust to off-target and specific on-target effects; thus they may be used to assess the shared on-target effect on a gene by multiple siRNAs. Different siRNAs may contribute to the shared effects with different weights. Considering the fact that the contributing weights are hardly known, here we concentrate on the cases in which each of m siRNAs against a gene has an equal weight of $1/m$. (See Section 8.7 of Chapter 8 for how to handle cases in which different siRNAs have different contributing weights.)

To calculate collective SSMD for multiple siRNAs, we must explore sources of data variation. The variation of data for siRNAs targeting a gene may come from two sources: (i) the variation among different siRNAs (i.e., the variation of the black crosses in a panel in Figure 6.5), and (ii) the variation among replicates within an siRNA (i.e., the variation of the bars next to each other with the same color for an siRNA in a panel in Figure 6.5). The former is contributed by different effects of siRNAs with different sequences matched the same gene. If we treat an siRNA as a biological replicate of a gene, then we may call this variation biological

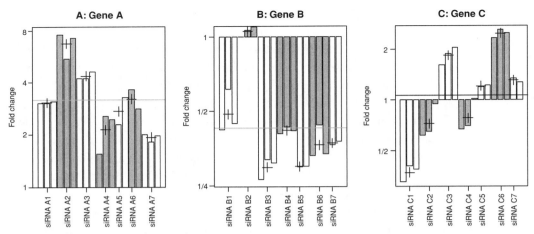

Figure 6.5   Examples of three genes, each targeted by seven siRNAs with different sequences in a deconvolution screen. In each panel, a bar represents the fold change in one replicate of an siRNA, the bars next to each other with the same color represent the values of replicates for one labeled siRNA, a black cross represents the mean of replicates for an siRNA, and a gray horizontal line represents the mean of all seven siRNAs against a gene.

variation. This variation may not be reduced or eliminated by improving assay quality. On the other hand, the variation of replicates of an siRNA is mostly produced in the experimental process and thus can be reduced or minimized by improving assay quality. Consequently, we may call it technical variation. The SSMD based on biological variation for multiple siRNAs is called biological collective SSMD, the SSMD based on technical variation is called technical collective SSMD, and the SSMD based on both variations is called total collective SSMD. The formulas for calculating average fold change, biological, technical, and total collective SSMD are shown in Table 6.3A. More details about how to derive these formulas are provided in Section 8.7 of Chapter 8.

In a deconvolution screen with good quality, the biological variation is usually larger than the technical variation. The existence of off-target effects in one or more siRNAs targeting the same gene usually causes large biological variation for a gene. Thus, in addressing off-target effects, the biological variation should be a more serious concern than the technical variation. Consequently, biological or total collective SSMD is usually more important than technical collective SSMD. The biological collective SSMDs are 2.9, −2.1, and 0.1, and the average fold changes are 3.2, 1/2.3, and 1.06, respectively, for siRNAs against genes A through C in Figure 6.5, indicating that siRNAs against gene A lead to a strong up-regulated effect (Figure 6.5A), siRNAs against gene B lead to a strong down-regulated effect (Figure 6.5B), and siRNAs against gene C barely lead to an extremely weak effect (Figure 6.5C).

Similar to classic t-statistic, the SSMD method may over-emphasize the genes with consistent but weak effects for its targeting siRNAs. Considering that the mean deviation may have to have a reasonable size, we may use dual-flashlight plot [170]

Table 6.3. Collective activity of multiple siRNAs with different sequences against a target gene and their individual and specific activities

---

A. Collective Activity of m siRNAs against a Gene

Mean $\qquad$ $\hat{\mu}_{\text{collective}} = \bar{d}_{\bullet\bullet} = \frac{1}{m} \sum_{i=1}^{m} \bar{d}_{i\bullet} = \frac{1}{m} \sum_{i=1}^{m} \left(\frac{1}{n_i} \sum_{j=1}^{n_i} d_{ij} \right)$

SSMD $\quad$ Biological $\quad$ $\hat{\beta}_{\text{biological}} = \dfrac{\hat{\mu}_{\text{collective}}}{\sqrt{\text{MSB}}}$

$\qquad\qquad$ Technical $\quad$ $\hat{\beta}_{\text{technical}} = \dfrac{\hat{\mu}_{\text{collective}}}{\sqrt{\text{MSE}}}$

$\qquad\qquad$ Total $\qquad$ $\hat{\beta}_{\text{total}} = \dfrac{\hat{\mu}_{\text{collective}}}{\sqrt{\text{MSB} + \text{MSE}}}$

B. Individual Activity of the ith siRNA

Mean $\qquad\qquad\qquad$ $\hat{\mu}_i = \bar{d}_{i\bullet}$

SSMD $\qquad\qquad$ $\hat{\beta}_i = \dfrac{\bar{d}_{i\bullet}}{s_i}$ or $\hat{\beta}_i = \dfrac{\Gamma\left(\dfrac{n_i - 1}{2}\right)}{\Gamma\left(\dfrac{n_i - 2}{2}\right)} \sqrt{\dfrac{2}{n_i - 1}} \dfrac{\bar{d}_{i\bullet}}{s_i},$

$\qquad\qquad\qquad\qquad\qquad\qquad\qquad$ where $\Gamma\left(\cdot\right)$ is a gamma function

C. Specific Activity of the ith siRNA

Mean $\qquad\qquad\qquad$ $\hat{\tau}_i = \bar{d}_{i\bullet} - \bar{d}_{\bullet\bullet}$

SSMD $\qquad\qquad$ $\hat{\lambda}_i = \dfrac{\hat{\tau}_i}{\sqrt{\text{MSE} \cdot \sum_{k=1}^{m} c_k^2}}$, where $c_k = \begin{cases} 1 - \dfrac{1}{m}, & \text{when } k = i \\[2mm] -\dfrac{1}{m}, & \text{when } k \neq i \end{cases}$

Note: Among the m siRNAs that are measured separately against a gene, the ith siRNA has been measured n_i times (i.e., measured in n_i different plates) and has a sample mean $\bar{d}_{i\bullet}$, and sample variance s_i^2. d_{ij} $(j = 1, \ldots, n_i)$ is the difference between the measured value of this siRNA and the mean or median value of a negative reference in a plate calculated in a way similar to what is described in Section 5.3.1. $\text{MSB} = \frac{1}{m} \sum_{i=1}^{m} (\bar{d}_{i\bullet} - \bar{d}_{\bullet\bullet})^2$, $\text{MSE} = \frac{1}{N-m} \sum_{i=1}^{m} \sum_{j=1}^{n_i} (d_{ij} - \bar{d}_{i\bullet})^2$, $N = \sum_{i}^{m} n_i$ and $\bar{d}_{\bullet\bullet} = \sum_{i=1}^{m} \bar{d}_{i\bullet}$. More details about how to derive these formulas are described in Section 8.7 of Chapter 8.

to display both average fold change and SSMD for the shared on-target effect of siRNAs against each gene and use the dual-flashlight method to select hits in the gene level.

6.6.4 SSMD-Based Methods Assessing Off-Target Effects and Specific On-Target Effects

In addition to assessing the collective activity of multiple siRNAs targeting a gene, we may be interested in the effect of each individual siRNA. The individual activity of an siRNA can be assessed based on its replicates using the sample mean $\bar{d}_{i\bullet}$ and SSMD $\hat{\beta}_i$, as shown in Table 6.3B. This activity of an siRNA contains the contributions from not only the shared on-target effect, but also the specific on-target and off-target effects. To evaluate off-target effects, we need to assess the specific effect of an siRNA relative to the collective effects of all siRNAs targeting the same gene. This specific effect can be assessed using mean and SSMD as shown in Table 6.3C. See Section 8.7 of Chapter 8 for mathematical details about how to derive the formulas in Table 6.3C.

The specific effect of an siRNA is a combination of off-target effect and specific on-target effect that this siRNA has. It is impossible to completely separate an off-target effect from a specific on-target effect in current designs of studies. However, the consideration of both specific and collective effects may provide a reference about which siRNA singles are more likely to have off-target effects and which are more likely to have specific on-target effects. The siRNAs with large specific effects for themselves but small collective effects for their targeted genes are more likely to have large off-target effects. The siRNAs with large specific effects in one direction but large collective effects in the opposite direction are also more likely to have large off-target effects. The siRNAs with large specific effects in one direction and large collective effects in the same direction are more likely to have large specific on-target effects.

6.7 Applications

6.7.1 An Example of an RNAi Confirmatory Screen with Replicates

Following the HCV RNAi primary screen in Section 5.5 of Chapter 5 and corresponding biological pathway analysis, a total of 640 siRNAs were selected and then arranged in a confirmatory screen in which each siRNA has three replicates. Based on Formula 6.2 using the negative control as the negative reference, we can estimate SSMD and mean difference for each siRNA as well as for each control well in a source plate. If the true value equals the estimated value of the SSMD, then the strength of effects for the siRNAs in the HCV RNAi confirmatory screen can be classified as in Table 6.4, which indicates 13 siRNAs with extremely strong, 22 with very strong, 22 with strong, 15 with fairly strong, and 38 with moderate inhibition effects.

However, the true value of SSMD is unknown. Thus the results in Table 6.4 are only approximations, especially when taking into consideration the small sample size of the screen. Consequently, we need to determine a critical value of SSMD for selecting hits so that desired control FPRs and FNRs may be achieved. Using the formulas in Table 6.1, we calculate p-values with respect to true SSMD values of 0 and -0.25 for addressing FPR and the p^*-values with respect to the true SSMD values of -2, -3, and -5 for addressing FNR in the direction of inhibition (Figure 6.6A). Based on the p-value and p^*-value, we can control FPR and FNR for each critical value of SSMD [169;177]. For example, the critical value of -1.28 controls FPR with respect to SSMD $= 0$ under 0.08 (black round points in Figure 6.6A) and FPR with respect to SSMD $= -0.25$ under 0.14 (black triangle points in Figure 6.6A) when

Table 6.4. The size of siRNA effects classified using SSMD in the HCV primary screen

Effect Subtype	Activation		Inhibition	
	SSMD Cutoff	Count	SSMD Cutoff	Count
Extremely strong	SSMD $\geq$ 5	4	SSMD $\leq$ −5	13
Very strong	5 > SSMD $\geq$ 3	4	−5 < SSMD $\leq$ −3	22
Strong	3 > SSMD $\geq$ 2	12	−3 < SSMD $\leq$ −2	22
Fairly strong	2 > SSMD $\geq$ 1.645	8	−2 < SSMD $\leq$ −1.645	15
Moderate	1.645 > SSMD $\geq$ 1.28	15	−1.645 < SSMD $\leq$ −1.28	28
Fairly moderate	1.28 > SSMD $\geq$ 1	11	−1.28 < SSMD $\leq$ −1	37
Fairly weak	1 > SSMD $\geq$ 0.75	30	−1 < SSMD $\leq$ −0.75	37
Weak	0.75 > SSMD > 0.5	38	−0.75 < SSMD < −0.5	64
Very weak	0.5 $\geq$ SSMD > 0.25	55	−0.5 $\leq$ SSMD < −0.25	70
Extremely weak	0.25 $\geq$ SSMD > 0	91	−0.25 $\leq$ SSMD < 0	64
Zero	SSMD = 0	0		

Note: "Count" is the number of siRNAs in each category.

we use the following decision rule: an siRNA is selected as a hit if it has estimated SSMD greater than −1.28 and as a non-hit otherwise in the direction of inhibition. This critical value also controls FNR with respect to SSMD = −3 under 0.018 (grey triangle points in Figure 6.6A) and FNR with respect to SSMD = −2 under 0.15 (grey round points in Figure 6.6A).

Given a decision rule for selecting hits, FPR is the proportion of all the true non-hits in a study selected as hits, and FNR is the proportion of all the true hits in the study declared as non-hits. The number of true non-hits in a study is unknown; so is the number of true hits in a study. What we know is the number of siRNAs declared as hits, as well as the number of siRNAs declared as non-hits. Therefore, FPR and FNR do not provide the exact information in which we are normally interested in an RNAi screen. We are primarily interested in the proportion of true non-hits among declared hits (i.e., FDR) and the proportion of true hits among declared non-hits (i.e., FNDR). The level of FDR is reflected by q-value, and the level of FNDR is reflected by q^*-value. Based on the p-values calculated using the formulas in Table 6.1, we can calculate corresponding q-value using R packages such as *qvalue* [146], *multtest* [11], and *fdrtool* [147]. Similarly, we can calculate q^*-value corresponding to p^*-value. The q-values and q^*-values calculated using *fdrtool* in the HCV confirmatory screen are shown in Figure 6.6B.

The critical value of −1.28 leads to the selection of 100 hits and 540 non-hits in the direction of inhibition. As described in Section 5.3.1 of Chapter 5, to define false positives and false negatives in the SSMD method, we need to specify two different values, β_1 and β_2, of true SSMD. In the case of $\beta_1 = -3$ and $\beta_2 = -0.25$, the q-value is 0.66 (black triangle points in Figure 6.6B), and the q^*-value is 0.003 (grey triangle points in Figure 6.6B) when we use the critical value of −1.28 for selecting inhibition

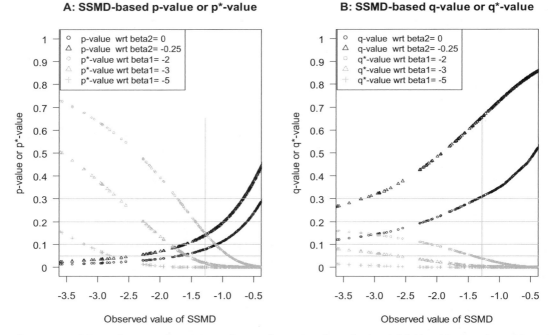

Figure 6.6   (A) *p*-value, *p**-value, (B) *q*-value, and *q**-value for selecting inhibition hits in a hepatitis C virus RNAi confirmatory screen with three replicates.

hits. That is, on average, there are no more than 66 (i.e., 100×0.66) false positives with respect to $\beta_2 = -0.25$ (i.e., siRNAs with extremely weak or no inhibition effects) among the 100 selected hits and no more than 2 (i.e., 540×0.003) false negatives with respect to $\beta_1 = -3$ (i.e., siRNAs with very strong or extremely strong effects) among the 540 declared non-hits in the inhibition direction. Similarly, in the case of $\beta_1 = -2$ and $\beta_2 = 0$, the *q*-value is 0.31 (black round points in Figure 6.6B) and the q^*-value is 0.04 (grey round points in Figure 6.6B) when we use the critical value of -1.28 for selecting inhibition hits. That is, on average, there are no more than 31 siRNAs with no inhibition effects among the 100 selected hits and no more than 22 siRNAs with strong, very strong, or extremely strong inhibition effects among the 540 declared non-hits.

As described in Section 6.2, due to the small sample size, a dual-flashlight plot (or volcano plot) will be useful in considering both SSMD (*p*-value or *q*-value) and mean difference (or equivalently average fold change) for selecting hits. Figure 6.7 indicates that the average fold changes for all the wells of the stronger positive control in the source plate are greater than two-fold (purple points in Figure 6.7) and those for the weaker positive control are all greater than 1.1-fold (black points in Figure 6.7) in the inhibition direction. Considering the fact that the weaker positive control is more biologically relevant, we use the cutoff of 1.1 for average fold change in this experiment.

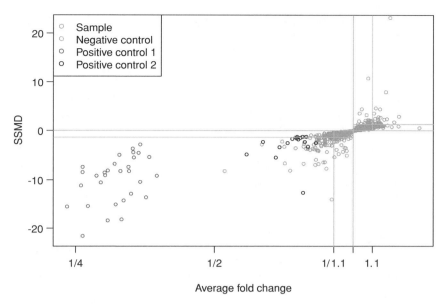

Figure 6.7   (See color insert following page 110.) Dual-flashlight plot for selecting inhibition hits in a hepatitis C virus RNAi confirmatory screen with three replicates. Positive control 1 and Positive control 2 are for the stronger and weaker positive controls, respectively. The horizontal gray lines denote SSMD $= -1.28$ and 1.28, respectively.

Using both SSMD and mean difference, we obtain the following result: 70 siRNAs with estimated SSMD values less than -1.28 and average fold change greater than 1.1-fold in the inhibition direction and 23 siRNAs with estimated SSMD values greater than 1.28 and average fold change greater than 1.1-fold in the activation direction (Figure 6.7). Those 93 siRNAs should be the main focus in follow-up research.

6.7.2 An Example of an RNAi Deconvolution Screen with Multiple Single siRNAs per Gene

In another project for cancer, following a confirmatory screen, 344 genes were selected for a further study in a deconvolution screen. Of these 344 genes, 343 genes were targeted by 7 single siRNAs and 1 gene by 10 siRNAs, resulting in the measurement of 2,411 single siRNAs. Each of these siRNAs had two to six replicates. We used the formulas in Table 6.3 to calculate average fold change and SSMD for each individual siRNA, average fold change and collective SSMD for each set of siRNAs targeting a gene, and specific activity for each siRNA beyond its shared activity.

The biological collective SSMD may address off-target effects better than technical and total collective SSMD. Consequently, we concentrate on the use of the biological collective SSMD. The classifying results based on the biological collective SSMD for siRNAs targeting a gene indicate that 2 genes have extremely strong effects, 4 strong, 5 fairly strong, 6 moderate, 19 fairly moderate, 22 fairly weak, 40 weak, 57 very weak, and 36 extremely weak in the inhibition direction, as well as 6 strong, 2 fairly

Table 6.5. The size of gene effects based on SSMD in a cancer deconvolution screen

Effect Subtype	Activation		Inhibition	
	SSMD Cutoff	Count	SSMD Cutoff	Count
Extremely strong	$SSMD \geq 5$	0	$SSMD \leq -5$	2
Very strong	$5 > SSMD \geq 3$	0	$-5 < SSMD \leq -3$	0
Strong	$3 > SSMD \geq 2$	6	$-3 < SSMD \leq -2$	4
Fairly strong	$2 > SSMD \geq 1.645$	2	$-2 < SSMD \leq -1.645$	5
Moderate	$1.645 > SSMD \geq 1.28$	7	$-1.645 < SSMD \leq -1.28$	6
Fairly moderate	$1.28 > SSMD \geq 1$	10	$-1.28 < SSMD \leq -1$	19
Fairly weak	$1 > SSMD \geq 0.75$	25	$-1 < SSMD \leq -0.75$	22
Weak	$0.75 > SSMD > 0.5$	25	$-0.75 < SSMD < -0.5$	40
Very weak	$0.5 \geq SSMD > 0.25$	33	$-0.5 \leq SSMD < -0.25$	57
Extremely weak	$0.25 \geq SSMD > 0$	45	$-0.25 \leq SSMD < 0$	36
Zero	$SSMD = 0$	0		

strong, 7 moderate, 10 fairly moderate, 25 fairly weak, 25 weak, 33 very weak, and 45 extremely weak in the activation direction (Table 6.5).

We used the dual-flashlight plot in Figure 6.8, which incorporates both average fold change and biological collective SSMD, to select hits at the gene level. The cutoffs of 1/1.2 for average fold change and −0.5 for biological collective SSMD lead to the selection of 97 genes with at least weak inhibition effects and at least 1.2-fold change on average in the inhibition direction, as well as 72 genes with at least weak activation effects and at least 1.2-fold change on average in the activation direction. The cutoffs of 1/1.5 for average fold change and −1 for biological collective SSMD lead to the selection of 31 genes with at least fairly moderate inhibition effects and at least 1.5-fold change in the inhibition direction and 20 genes with at least fairly moderate activation effects and at least 1.5-fold change in the activation direction. Those genes, especially the 51 genes with absolute value of SSMD greater than 1 and average fold change greater than 1.5 in either direction, should be explored in follow-up researches.

The specific effect of each siRNA beyond its shared on-target effects was evaluated using both average log fold change and SSMD, calculated using formulas in Table 6.3C. The specific effects of siRNAs in the study can be evaluated using dual-flashlight plot, as shown in Figure 6.9. From Figure 6.9, 423 of 2,411 siRNAs have absolute SSMD value greater than 1.28 and average fold change greater than 2 in either direction. Those siRNAs may have a large off-target effect, a large specific on-target effect, or both. We may further check those 423 siRNAs in a scatter plot of average fold change of a single siRNA versus that of multiple siRNAs against its targeting gene, as shown in Figure 6.10.

The specific effect of an siRNA is a combination of off-target effect and specific on-target effect that this siRNA has. It is impossible to separate an off-target effect

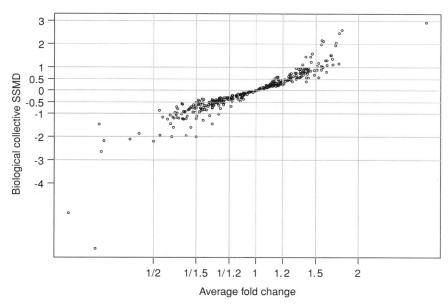

Figure 6.8   Dual-flashlight plot for selecting hits based on the collective activity of multiple siRNAs tar-geting each gene in a cancer siRNA deconvolution screen.

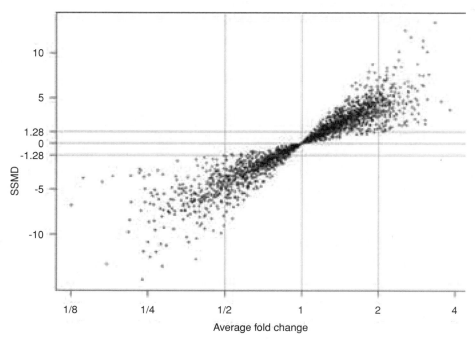

Figure 6.9   Dual-flashlight plot for the specific effect consisting of combined off-target effects and specific on-target effects for each siRNA in a screen. One point denotes the estimated SSMD value and average fold change for the specific effect of a single siRNA.

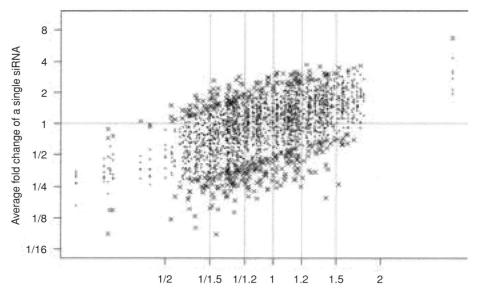

Figure 6.10 Scatter plot displaying the average fold change in an siRNA versus that in its targeting gene. Black "×" points denote siRNAs with large specific effects (i.e., SSMD > 1.28 and average fold change > 2-fold in either direction).

from a specific on-target effect in the calculated values in current study designs. However, Figure 6.10 can provide a reference regarding which siRNAs are more likely to have off-target effects and which siRNAs are more likely to have specific on-target effects among the 423 siRNAs with large specific effects ("×" points).

The "×" points for the siRNAs whose targeting genes have average fold change away from 1 (especially the 83 siRNAs marked with "×" points whose targeting genes have average fold change less than 1/1.5 or greater than 1.5) are more likely to have large specific on-target effects (Figure 6.10). Two of them are shown in Figure 6.11A and B. The average fold change and biological collective SSMD are, respectively, 3.17 and 2.92 for siRNAs A1 through A7 targeting gene A (Figure 6.11A). Thus these siRNAs have a large shared on-target activation effect. siRNA A2 has an average fold change of 6.7, whereas the remaining 6 siRNAs all have an average fold change of approximately 3. The average fold change and SSMD for the specific activity of siRNA A2 are 2.1 and 4.6, respectively. Thus siRNA A2 against gene A has a large specific on-target activation effect. Similarly, the average fold change and biological collective SSMD are, respectively, 1/2.86 and −2.63 for siRNAs B1 and B2 targeting gene B (Figure 6.11B). Thus these siRNAs against gene B have a large shared on-target inhibition effect. siRNA B5 has an average fold change of 1/6.7, whereas the remaining 6 siRNAs all have an average fold change of approximately 1/2.5. The average fold change and SSMD for the specific activity of siRNA B5 are 1/2.36 and

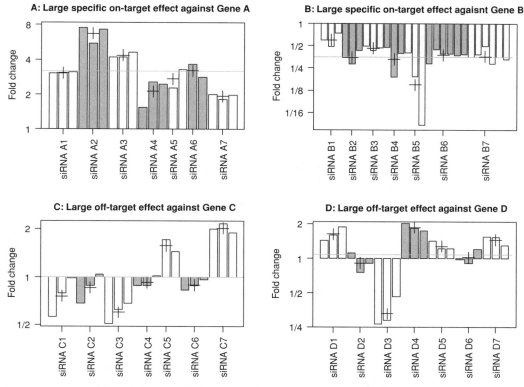

Figure 6.11   Examples of siRNAs with large specific on-target effects (A, siRNA A2; B, siRNA B5) or with large off-target effects (C, siRNA C7; D, siRNA D5). In each panel, a bar represents the fold change in one replicate of an siRNA, the bars next to each other with the same color represent the values of replicates for one labeled siRNA, a black cross represents the mean of replicates for an siRNA, and a gray horizontal line represents the mean of all seven siRNAs for a gene.

−2.11, respectively. Thus siRNA B5 against gene B has a large specific on-target inhibition effect.

The "×" points with average fold change of approximately 1 (especially the 164 siRNAs in red with average fold change between 1/1.2 and 1.2) are more likely to have large off-target effects (Figure 6.10). This is because those siRNAs have large effects, whereas their targeting genes do not. Two of these siRNAs are shown in Figures 6.11C and D. The average fold changes and biological collective SSMDs are, respectively, 1.004 and 0.01 for siRNAs C1 through C7 targeting gene C (Figure 6.11C). Thus these siRNAs have essentially no shared on-target effect. siRNA C7 has an average fold change of 2.02, whereas the remaining six siRNAs all have an average fold change of approximately 1. The average fold change and SSMD for the specific activity of siRNA C7 are 2.01 and 4.3, respectively. Thus siRNA C7 against gene C has a large off-target activation effect. Similarly, the average fold change and biological collective SSMD are, respectively, 1.08 and 0.14 for siRNAs D1 through D7 targeting gene D (Figure 6.11D). Thus these siRNAs against gene D have a very small shared on-target effect. siRNA D3 has an average fold change of 1/3.04, whereas the remaining six siRNAs have average fold change of approximately 1. The average fold change and SSMD for the specific activity of siRNA D3 are 1/3.3 and −7.3, respectively. Thus siRNA D3 against gene D has a large off-target inhibition effect.

6.8 Discussion and Conclusions

The ultimate goal in a genome-scale RNAi project is to select siRNAs with a desired size of inhibition or activation effects on a biological phenotype of interest. Two main strategies are commonly used in hit selection in RNAi screens: (i) ranking, and (ii) testing [161].

Traditionally, two types of measures have been used to rank siRNA effects. One is mean difference (or, equivalently, average fold change), and the other is the p-value from either the z-score method or a t-test of mean difference. The first measure cannot represent the magnitude of the difference because it does not effectively capture the data variability [162;165]. When statistical significance is used, the p-value comes from testing the hypothesis of no mean difference between two groups. It addresses the question of whether an siRNA has exactly the same effect as the negative reference based on the sample observation. It is also not designed to measure the magnitude of the difference is [162;165]. The p-value from a z-score or t-test is affected by both sample size and the size of the siRNA effect. An siRNA effect that results in a low p-value may not cause a robust enough effect on the assay to indicate any meaningful biological association. Therefore, neither mean difference nor p-value can represent the magnitude of difference.

SSMD is a better metric for measuring the magnitude of difference [162]. SSMD can be calculated as the mean of log fold change divided by standard deviation of log fold change with respect to a negative reference. Thus SSMD can be interpreted as the average fold change (in log scale) penalized by the variability of fold change (in log scale). Unlike mean difference, SSMD is robust to both measurement unit and strength of positive controls; it takes into account data variability in both compared groups and has a probability interpretation [166;167]. Compared with p-value, SSMD directly measures how large the magnitude of difference is [162;165;167]. Therefore, SSMD effectively measures the size of siRNA effects. SSMD-based classifying rules (Table 5.2 of Chapter 5) can be used to assess the strength of siRNA effects in RNAi screens [167].

In the testing strategy for screens with replicates, a t-test is popularly used. Traditionally, the t-test for testing no mean difference controlled the error rate in which we conclude that there exists a (possibly tiny) mean difference, when actually there is no mean difference. However, it is well known that cells are controlled by dynamic actions of thousands of genes that are related through complex interaction. Because of the existence of gene networks, an siRNA will rarely have exactly the same effect as the negative reference. Thus the null hypothesis of no mean difference is unrealistic for many genes. As a remedy, we may consider testing a null hypothesis of mean difference using a constant other than zero. However, because mean difference may have different meanings in different experiments and because mean difference cannot effectively assess the size of siRNA effects, it is hardly feasible to set up this constant for every experiment.

As with testing mean difference, we control the error rate, testing whether the null hypothesis of SSMD is zero. The thresholds of SSMD are applicable to any experiment; consequently, it is feasible to test whether SSMD is a value other than zero (such as 0.25 or 3) [178]. Using SSMD, we can maintain balanced control of FPRs, in which siRNAs with no or extremely weak effects (e.g., SSMD ≤ 0.25) are selected as hits, and FNRs, in which the siRNAs with very strong effects (e.g., SSMD ≥ 3) are not selected as hits. To adjust for multiplicity issues, we can construct corresponding q-value and q^*-value based on SSMD to address FDRs and FNDRs in RNAi screens.

Off-target effects offer a challenge in genome-scale RNAi screens [78]. Although the ideal approach of saving experiments from off-target effects remains technically challenging in most contexts, a solution, adopted by most researchers and companies today, is to test multiple siRNAs with different sequences against a target gene to increase the level of confidence in positive hits [14;40]. Consequently, hit selection accounting for off-target effects is commonly achieved through the deconvolution screen in which multiple independently active siRNAs that target the same gene are measured in an experiment. The SSMD-based method in Section 6.6 captures the collective activity of siRNAs targeting a gene. It naturally incorporates all the information of multiple siRNAs targeting the same gene in a strong statistical basis. Consequently, the results are reliable. In addition, this method can assess not only the activity of a gene targeted by multiple siRNAs, but also the strength of specific effect (consisting of specific on-target effect and off-target effect) for each siRNA beyond the shared on-target effect of siRNAs on their targeting gene. The consideration of effects at both the gene level and the siRNA level can also give a reference about which siRNAs are more likely to have large off-target effects and which siRNAs are more likely to have specific on-target effects, as shown in Figures 6.10 and 6.11.

A typical genome-scale RNAi project contains at least two of the following three types of screens: (i) a primary screen with or without replicates, (ii) a confirmatory screen with replicates, and (iii) a deconvolution screen with multiple siRNAs per gene. Many methods can be applied to only one type of screen. For example, the commonly used z^*-score method is only applicable to a primary screen without replicates. The classical t-test is only applicable to screens with replicates. The frequency approach for addressing off-target effects is only applicable in deconvolution screens. The RSA and RIGER methods is inapplicable in deconvolution screens where most investigated siRNAs are preselected to have non-zero effects in one direction.

SSMD is applicable to screens without replicates, screens with replicates, and any deconvolution screens. Moreover, SSMD has strong theoretical bases and produces more reasonable results than other methods, as shown in this chapter and Chapter 5. Therefore, the SSMD method, along with the dual-flashlight plot, should be the first choice for hit selection in genome-scale RNAi screens.

Part II

Methodological Development for Analyzing RNAi HTS Screens

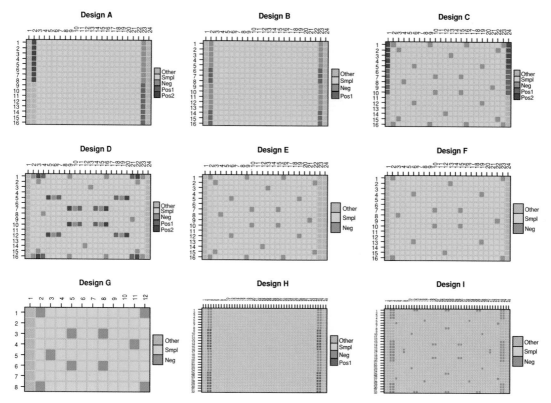

Plate 2.1   Plate designs in a 384-well (designs A–F), 96-well (design G), or 1,536-well plate (designs H and I). Designs A, B, and H are for situations in which controls can only be arranged in edge columns; designs C through G and I are for situations in which controls are allowed to be arranged anywhere in a plate. The colors represent types of wells, as shown in the legend to the right of each panel: green = negative control; red = first positive control; purple = second positive control; yellow = sample siRNAs; gray = other controls. *Source:* From Zhang [166].

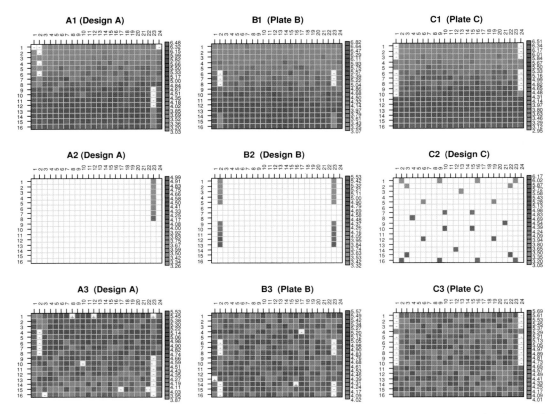

Plate 2.2   The capacity of three typical plate designs, designs A, B, and C (shown in Figure 2.1), in identifying and adjusting for linear row effects. Panels A1, B1, and C1 show the measured intensities (in log10 scale) in all wells in a plate from three experiments that have plate designs A, B, and C, respectively. Panels A2, B2, and C2 display intensities of the negative control. Panels A3, B3, and C3 display the data adjusted using the negative control wells. In each panel, a red + (or a green −) denotes an outlier in up-regulated (or down-regulated) direction based on sample wells.

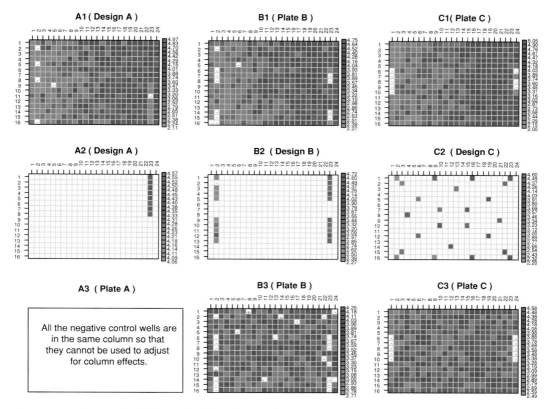

Plate 2.3   The capacity of three typical plate designs, designs A, B, and C (shown in Figure 2.1), in identifying and adjusting for linear column effects. Panels A1, B1, and C1 show the measured intensities (in log10 scale) in all wells in a plate from three experiments that have plate designs A, B, and C, respectively. Panels A2, B2, and C2 display intensities of the negative control. Panels A3, B3, and C3 display the data adjusted using the negative control wells. In each panel, a red + (or a green −) denotes an outlier in up-regulated (or down-regulated) direction based on sample wells.

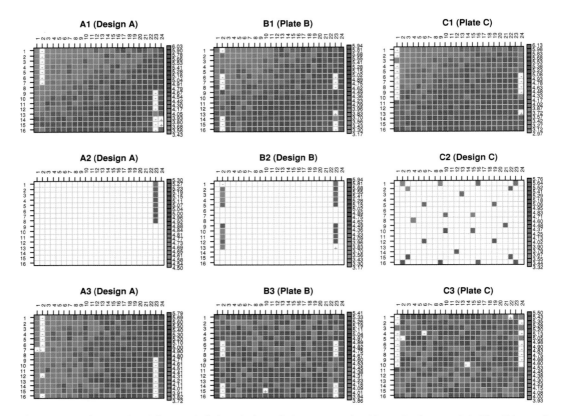

Plate 2.4   The capacity of three typical plate designs, designs A, B, and C (shown in Figure 2.1), in identifying and adjusting for linear row and column effects. Panels A1, B1, and C1 show the measured intensities (in log10 scale) in all wells in a plate from three experiments that have plate designs A, B, and C, respectively. Panels A2, B2, and C2 display intensities of the negative control. Panels A3, B3, and C3 display the data adjusted using the negative control wells. In each panel, a red + (or a green −) denotes an outlier in up-regulated (or down-regulated) direction based on sample wells.

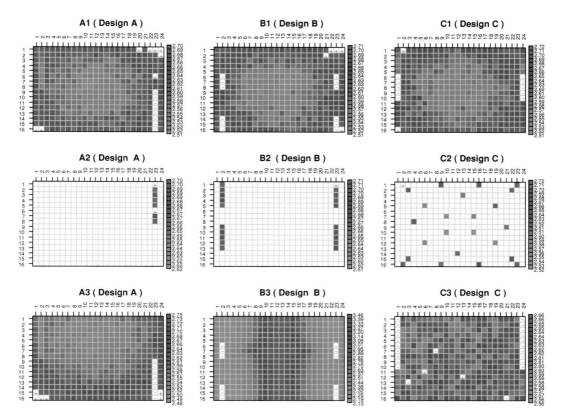

Plate 2.5   The capacity of three typical plate designs, designs A, B, and C (shown in Figure 2.1), in identifying and adjusting for bowl-shaped spatial effects. Panels A1, B1, and C1 show the measured intensities (in log10 scale) in all wells in a plate from three experiments that have plate designs A, B, and C, respectively. Panels A2, B2, and C2 display intensities of the negative control. Panels A3, B3, and C3 display the data adjusted using the negative control wells. In each panel, a red + (or a green −) denotes an outlier in up-regulated (or down-regulated) direction based on sample wells. *Source:* From Zhang [166].

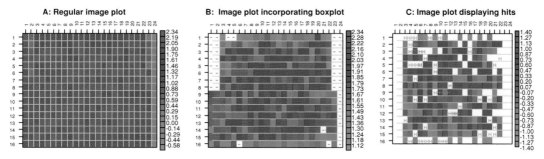

Plate 3.2   Regular image plot (Panel A), improved image plot (Panel B) to display the measured value in a plate, and improved image plot to display selected hits in a source plate (Panel C) in the HCV RNAi second confirmatory screen described in Section 1.4 in Chapter 1. The improved image plot clearly reveals a systematic spatial effect (low values in the edge rows and high values in the middle rows), whereas the regular image plot does not reveal the spatial effect.

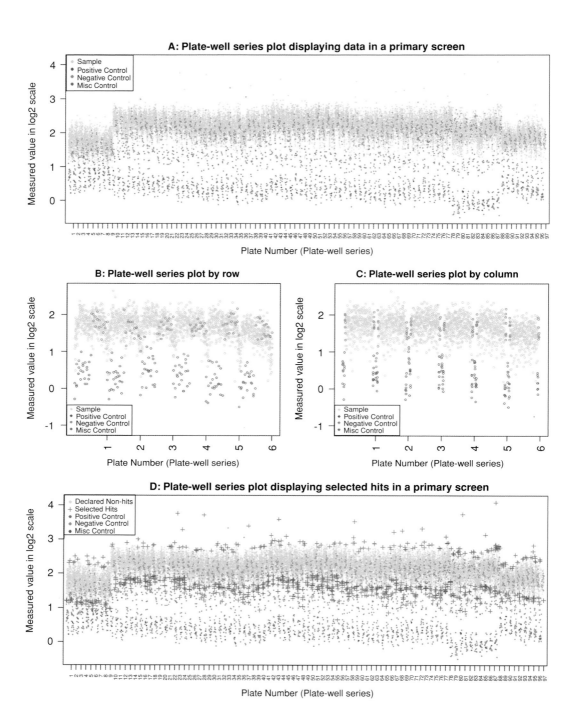

Plate 3.1   Plate-well series plots to display data or hits in the HCV RNAi primary screen (A, D) and the second confirmatory screen (B, C) described in Section 1.4. In the legends, Background, Positive Control, and Misc Control denote empty wells, stronger inhibition control wells, and weaker inhibition control wells, respectively.

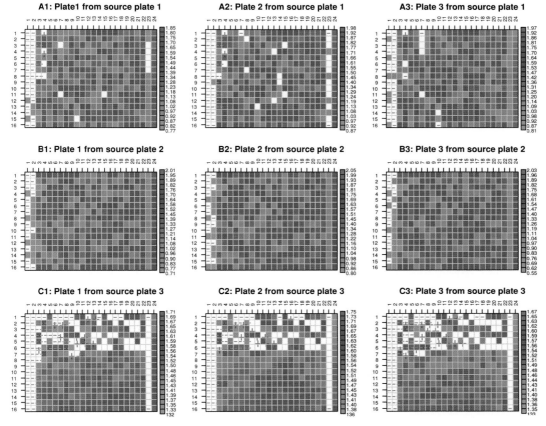

Plate 3.3   Improved image plots to display a high reproducibility of data in plates from three source plates (source plates 1, 2, and 3) in an RNAi confirmatory screen. The data in the plates from the same source plate have a very similar pattern, whereas the data in the plates from different source plates have very different patterns.

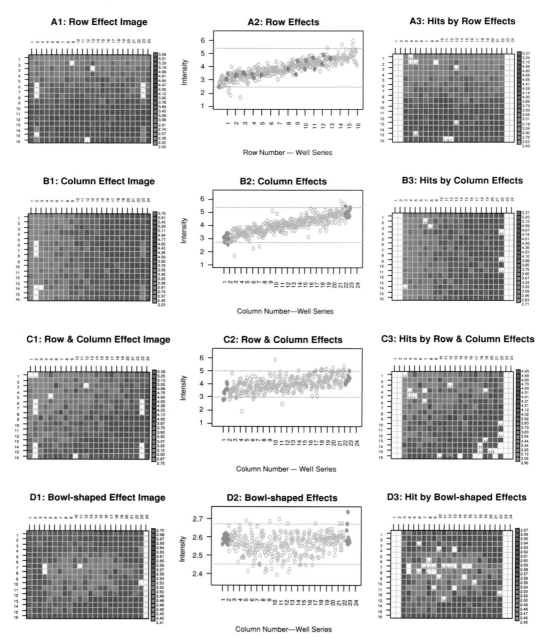

Plate 3.6   The use of improved image plot and plate-well series plot to display four common patterns of spatial effects: (i) linear row effects (A1–A3), (ii) linear column effects (B1–B3), (iii) linear row and column effects (C1–C3), and (iv) bowl-shaped spatial effects (D1–D3). In A1, B1, and C1, a red "+" (or a green "–") denotes an outlier in up-regulated (or down-regulated) direction. In A2, B2, and C2, an orange (or green) point denotes a value in a sample well (or a negative control well); the grey lines denote the boundary of SSMD = ±1.4 for selecting hits. In A3, B3, and C3, a red (or green) "H" denotes a selected hit in up-regulated (or down-regulated) direction.

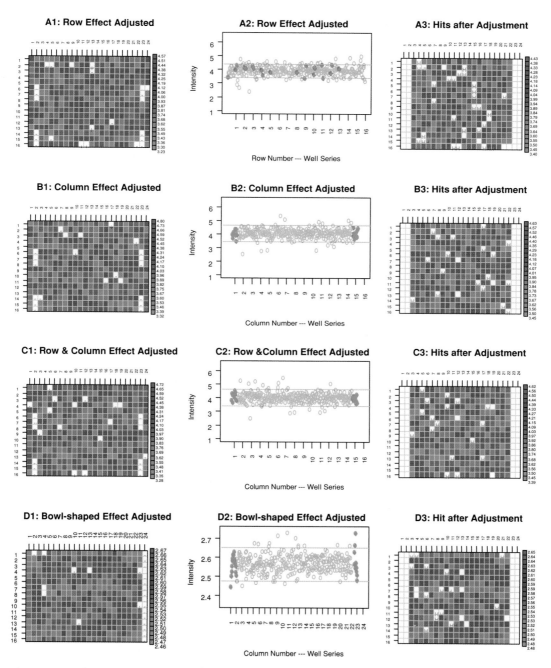

Plate 3.7   Adjustment of four common patterns of spatial effects: (i) linear row effects (A1–A3), (ii) linear column effects (B1–B3), (iii) linear row and column effects (C1–C3), and (iv) bowl-shaped spatial effects (D1–D3). In A1, B1, and C1, a red "+" (or a green "−") denotes an outlier in up-regulated (or down-regulated) direction. In A2, B2, and C2, an orange (or green) point denotes a value in a sample well (or a negative control well); the grey lines denote the boundary of SSMD = ±1.4 for selecting hits. In A3, B3, and C3, a red (or green) "H" denotes a selected hit in up-regulated (or down-regulated) direction.

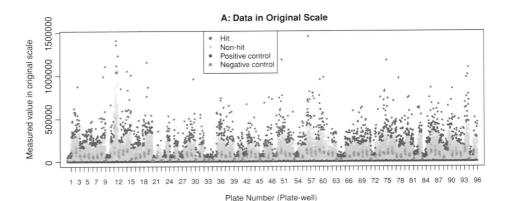

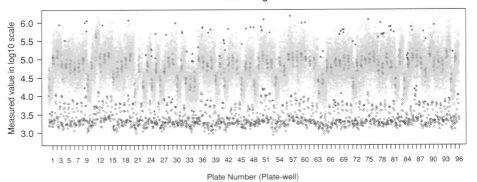

Plate 3.5   Plate-well series plots to display the data and selected hits based on original scale (A) or log-transformed scale (B). The use of the criterion of selecting the sample siRNAs with absolute z^*-score (robust version of z-score) greater than 3 as hits leads to the selection of 1,843 up-regulated hits and zero down-regulated hits based on the raw data (A, red points), but selection of 77 up-regulated hits and 29 down-regulated hits based on the log-transformed data (B, red points).

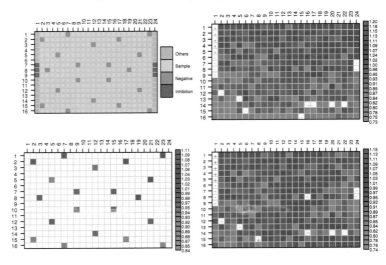

Plate 3.8   Adjustment of spatial effects using an effective plate design in a confirmatory screen: Top left: plate design; top right: raw data in a plate. Bottom left: raw data in negative control wells; bottom right: adjusted data in a plate.

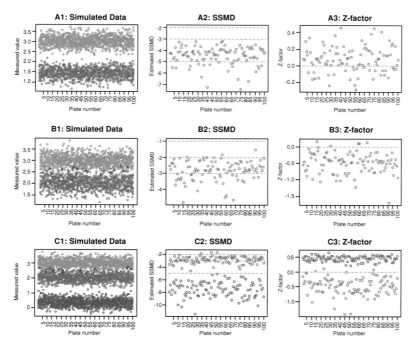

Plate 4.3   Data, SSMD, and Z-factor in three simulated experiments A, B, and C in which the positive controls have different effect sizes: a very strong control (red points) in experiment A, a strong control (red points) in experiment B, and an extremely strong control (purple points) and a strong control (red points) in experiment C. In each simulated experiment, there are 100 plates each, with 10 replicates for each positive or negative control. The data for each control in each experiment is generated from a normal distribution with standard deviation of 0.25, namely $N(\mu, 0.25^2)$ where $\mu = 3$ for the negative control in each experiment, and $\mu = 1.44, 2.01, 0.35, 2.01$ for the very strong positive control in experiment A, strong positive control in experiment B, and extremely strong positive control and strong positive control in experiment C, respectively.

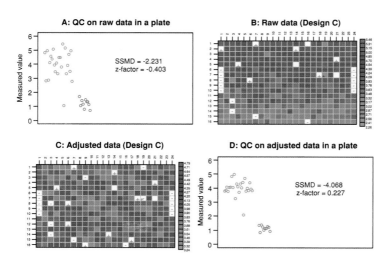

Plate 4.5   An effective plate design saving an assay that might have been judged as poor quality. The screen had strong row effects and adopted design C as shown in Figure 2.1 of Chapter 2. The red and green points respectively denote a very strong positive control and a negative control in A and D.

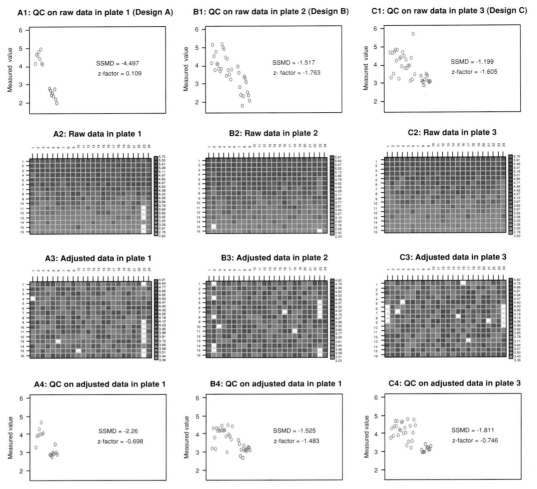

Plate 4.4   Impact of plate design on quality assessment. Three designs (A, B, and C as shown in Figure 2.1 of Chapter 2) are adopted in plates 1, 2, and 3, respectively, in a primary screen with strong row effects. The red and green points respectively denote a very strong positive control and a negative control in A1, A4, B1, B4, C1, and C4.

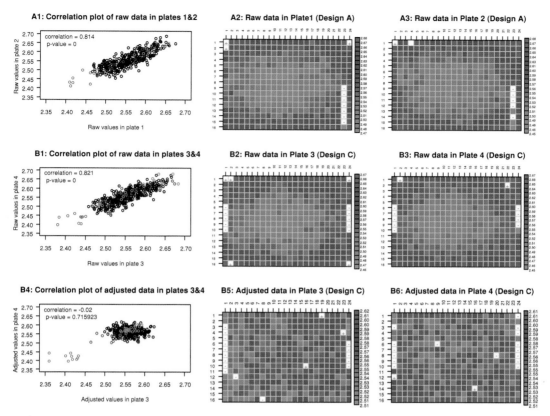

Plate 4.6   Impact of systematic spatial effects on quality assessment using correlation between two replicate plates from the same source plate. The red, blue, green, and black points respectively denote a very strong positive control, a moderate positive control, a negative control, and sample siRNA wells in A1, B1, and B4.

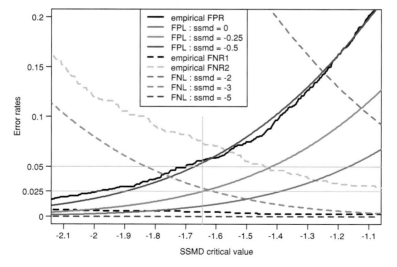

Plate 5.5   False-positive levels (FPLs), false-negative levels (FNLs), and empirical false-negative rates (FNRs) for selecting inhibition hits in a hepatitis C virus RNAi primary screen without replicates.

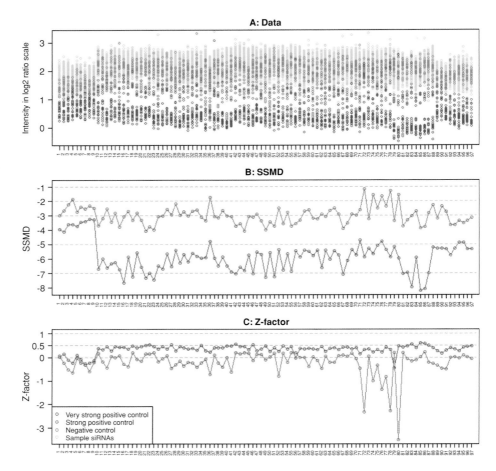

Plate 4.7   Quality assessment in an HCV siRNA primary experiment. The *x*-axis in each panel denotes plate numbers. A point denotes the measured intensity in a well of a plate in A, a value of SSMD in a plate in B, and a value of *Z*-factor in a plate in C. The well types are denoted using different colors, as shown in the legend of C.

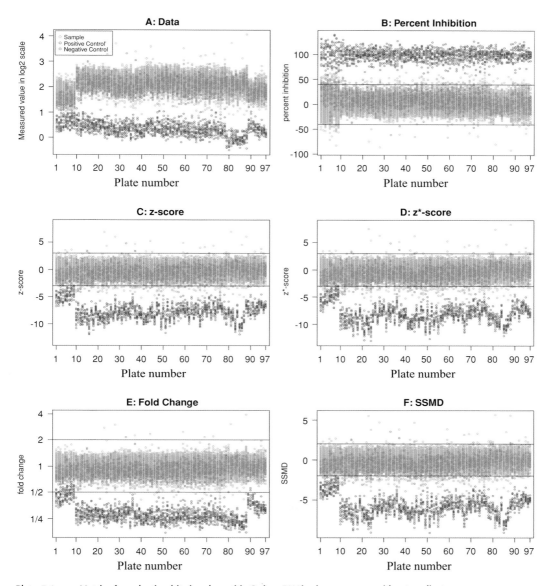

Plate 5.4   Metrics for selecting hits in a hepatitis C virus RNAi primary screen without replicates.

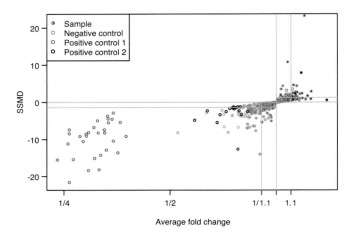

Plate 6.7   Dual-flashlight plot for selecting inhibition hits in a hepatitis C virus RNAi confirmatory screen with three replicates. Positive control 1 and Positive control 2 are for the stronger and weaker positive controls, respectively. The horizontal gray lines denote SSMD = −1.28 and 1.28, respectively.

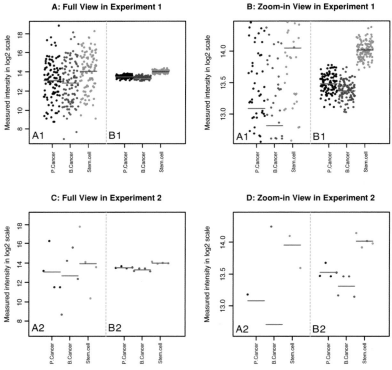

Plate 7.3   The measured intensity in log2 scale of three cell lines, namely prostate cancer (P.Cancer), breast cancer (B.Cancer), and stem cell lines (Stem.cell), for siRNAs A1 and B1 in simulated experiment 1 and for siRNAs A2 and B2 in experiment 2. The measured values in the three cell lines are from normal distributions $N(13.06, 2.04^2)$, $N(12.94, 2.04^2)$, and $N(14, 2.04^2)$, respectively, for siRNAs A1 and A2, and are from $N(13.48, 0.16^2)$, $N(13.35, 0.16^2)$, and $N(14, 0.16^2)$, respectively, for siRNAs B1 and B2. The number of replicates in each cell line is 100 in experiment 1 and 4 in experiment 2. A red segment denotes a sample mean of measured value in a cell line for an siRNA. The two cancer cell lines have an average two-fold decrease in original scale compared with the stem cell line for siRNAs A1 and A2 and an average 1.5-fold decrease for siRNAs B1 and B2. *Source:* From Zhang [167].

7

Statistical Methods for Group Comparison

In genome-scale RNAi screens, the primary objective is to select siRNAs with desired effect sizes, which relies on the comparison of gene effects in multiple different groups. Thus statistical methods for group comparisons play a critical role in data analysis in RNAi screens. A major statistical method for group comparison is contrast analysis. Traditionally, a contrast is a linear combination of group means in which the coefficients sum to zero. A typical contrast analysis is the significance testing of whether a contrast is zero. However, there are many issues with such contrast analysis. In fact, issues with the significance testing of a simple contrast (i.e., testing no mean difference between two groups) have incurred continuous calls for a critical reexamination of the common use of *null hypothesis significance testing* (NHST) in behavioral and social science [4;26;34;59;60;105;106;119;127;131;132], which has even led some researchers to advocate that the use of significance tests be banned in research [26;34;127;131;132]. The major issues with traditional contrast analysis are discussed in Section 7.1. We face similar issues when we apply traditional methods for group comparison to analyze data from genome-scale RNAi screens.

Recently, a new method of contrast analysis was proposed to address issues in traditional contrast [163;167;170]. This core of this new method is the concept of using a contrast variable, defined as a linear combination of random variables (with each variable representing random values in a group), instead of group means. This concept of contrast variable and two associated terms, *standardized mean of contrast variable* (SMCV) and c^+-probability, are critical for deriving statistical methods for assessing siRNA effects in genome-scale RNAi screens. Therefore, in this chapter, I present general concepts and theorems in the new contrast analysis. Specifically, I discuss the major issue of traditional contrast analysis in Section 7.1; present the general concepts of contrast variable, SMCV, and c^+-probability, along with a theorem to set up the relationships between SMD and c^+-probability, in Section 7.2; and describe a classifying rule for interpreting strength of group comparison in Section 7.3. The use of the new contrast relies on the estimation and inference of SMCV and c^+-probability. Thus I provide and prove a theorem to facilitate the

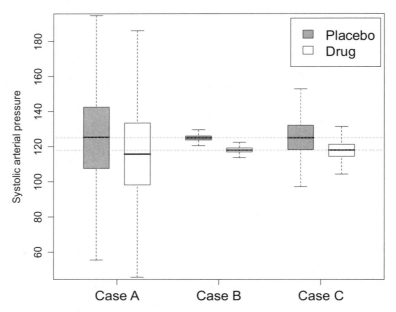

Figure 7.1   Boxplots to show the comparison of blood pressure of two groups treated with a placebo and a drug in three cases A, B, and C. The population distributions of the two groups are $N(125,25.5^2)$ and $N(116,26^2)$ in case A, $N(125,1.66^2)$ and $N(118, 1.66^2)$ in case B, and $N(125,10^2)$ and $N(118,5^2)$ in case C, respectively.

estimation and inference of SMCV in Section 7.4. Based on this theorem, I derive the estimation and inference of SMCV and c^+-probability in Section 7.5. Next I elaborate contrast variable, SMCV, and c^+-probability within the framework of multifactor analysis-of-variance (ANOVA) in Section 7.6, demonstrate the utility of contrast variable using case studies and simulation in Section 7.7, and finally present conclusions in Section 7.8.

7.1 Illustration of Issues in Traditional Contrast Analysis

NHST was pioneered by Karl Pearson [117] and was formally proposed by Ronald Fisher [49;50], with formulation from Jerzy Neyman and Egon Pearson [111;112] on power and type I and type II errors. Since then, NHST has become a very common statistical practice. However, the practice of NHST has not been without controversy. A major issue in the practice of NHST is that NHST is often used in situations in which scientific interest is neither in the question of average effect being exactly zero nor in the direction of average effects, but rather in the magnitude of effects. This is also the major issue in traditional contrast analysis, in which a significance test of zero contrast is often conducted in a situation in which scientists are more interested in knowing how large the contrast is than whether the contrast is zero [139].

One of the issues with traditional contrast analysis can briefly be illustrated in a two-group example about blood pressure in patients, shown in Figure 7.1. In this example, the interest is the comparison of blood pressure in two independent groups of patients treated with a placebo and a drug. A traditional contrast in this study is $L = \mu_1 - \mu_2$, where μ_1 and μ_2 are the population means in the two groups, respectively. Clearly, traditional contrast aims at the comparison of group means. Using population value of tradition contrast, we would conclude that the order of difference between the placebo and the drug in the three cases is $A > B = C$ because the population values of the traditional contrast in cases A, B, and C are 9, 7, and 7, respectively (Figure 7.1). The confidence interval of the traditional contrast and p-value of testing the traditional contrast being zero may lead to similar conclusions as long as the sample size is large enough. However, the population distributions clearly show that, by taking the drug, most patients do not get any benefit in case A, most patients get some benefit in case B, and some patients get some benefit in case C. This observation suggests that the order of difference between the placebo and the drug is $B > C > A$. Therefore, traditional contrast analysis does provide information about mean of difference, but it is not enough to address the difference in a distribution level.

Figure 7.1 clearly shows that, to effectively compare two groups, we need additional analysis to incorporate information in a distribution level. This may be achieved through direct comparison of distributions. In the case of two-group comparisons, hypothesis testing of two distributions in nonparametric analysis [68] and stochastic dominance [36] have been used for comparing two groups in a distribution level. Meanwhile, various probabilistic indices have also been proposed for distribution comparison of two groups [1;29;33;36;37;102;115;122;125;134;142;184]. The probabilistic indices are more or less related to one another. The hypothesis testing of two distributions is equivalent to analysis of confidence intervals of probabilistic indices. One probabilistic index is d^+-probability, which is the probability of a difference between two groups being greater than 0 [161;162;165]. An important feature of d^+-probability is that it can accommodate both independent and correlated groups, whereas most of the other probabilistic indices, such as $P(Y < X)$, are defined and estimated upon independent groups. The d^+-probability has been extended to c^+-probability for the comparison of not only two groups, but also more than two groups [163;167;170].

In addition to the direct comparison of distributions, a category of analytic methods for comparing groups beyond traditional contrast analysis of comparing group means is the use of certain parameters or statistics to capture both mean and variability of groups, which includes various effect sizes, similar to standardized mean differences [32;53;66;73;102;125;126;134;139;142;151;161;162]. However, different effect size measures are suitable for different types of data, and the interpretations of effect sizes are generally arbitrary and remain problematic even for the same effect size measure [114]. Currently, the commonly used standardized mean differences (e.g., Cohen's d, Glass's $\hat{\Delta}$, and Hedge's g) work only in the condition of independence and homoscedasticity. Recently, Zhang proposed strictly standardized mean difference (SSMD), which is the ratio of the mean and standard deviation of the difference between two groups and has been proposed as an effect size to compare two groups [161;162;165]. SSMD has been applied for data analysis in HTS biotechnologies [13;82;86;161–163;165;167–171;174;175;177;178;182]. An advantage of SSMD is that its values are comparable across different experiments. SSMD was then expanded to SMCV for comparing not only two groups, but also more than two groups [163;167;170]. Like SSMD, the values of SMCV are comparable across experiments. SMCVs are applicable in a group comparison context, with or without independence and with or without homoscedasticity.

In the past, effect size and traditional contrast seemed like two separate entities. Recently, they have been integrated for any comparison in contrast analysis, which is achieved through a definition of contrast variable: a contrast variable is defined as a linear combination of random variables (with each variable representing a group) instead of group means [163;167;170]. Consequently, an effect size (i.e., SMCV) and traditional contrast (i.e., contrast mean) are two characteristics of the same random variable (i.e., contrast variable). This concept of contrast variable and two associated terms, SMCV and c^+-probability, allow the use of SMCV and c^+-probability to effectively address questions regarding the strength of a comparison and the use of contrast mean to address questions regarding the comparison of group means, thus avoiding the misuse of traditional contrast (especially p-value of testing traditional contrast $= 0$) to address questions about the strength of a comparison. The definitions of contrast variable, SMCV, and c^+-probability are described in the next section.

7.2 Contrast Variable, SMCV, and c^+-Probability

Traditionally, a contrast has been defined as a linear combination of the means in which the coefficients sum to zero. Here, a contrast variable is defined as a linear combination of groups themselves, instead of group means, in which the coefficients sum to zero [163;167;170]. The mean of a contrast variable equals a traditional contrast. SMCV is the ratio of the mean and standard deviation of a contrast variable. Consequently, effect size (i.e., SMCV) and traditional contrast (i.e., contrast mean) are now two characteristics of the same random variable; thus contrast variable integrates both effect size and traditional contrast. c^+-probability is the probability that a contrast variable is positive. We may also define c^+-probability using replication probability as follows: if we get a random draw from each condition and calculate the sampled value of the contrast variable based on the random draws, then c^+-probability is the chance that the sampled values of the contrast variable are greater than 0 when the random draw process is repeated infinite times. c^+-probability is a probabilistic index accounting for distributions of compared groups, whereas SMCV is a variant of standardized mean difference (e.g., Cohen's d, Glass's $\hat{\Delta}$, and Hedge's g) incorporating both mean and variance of groups. There is a link between SMCV and c^+-probability; thus standardized mean difference and probabilistic index are integrated to effectively assess the strength of a comparison.

The concepts in contrast analysis can be formalized mathematically as follows. Suppose t groups $P_1, P_2, \ldots, P_t$ have means $\mu_1, \mu_2, \ldots, \mu_t$ and variances $\sigma_1^2, \sigma_2^2, \ldots, \sigma_t^2$, respectively. Then a traditional contrast L is $L = \sum_{i=1}^{t} c_i \mu_i$ where $\sum_{i=1}^{t} c_i = 0$. Traditional contrast analysis focuses on testing $H_0 : L = 0, H_0 : L \leq 0$ or $H_0 : L \geq 0$. A contrast variable V is $V = \sum_{i=1}^{t} c_i P_i$ where $\sum_{i=1}^{t} c_i = 0$. The SMCV of contrast variable V, denoted by λ, is $\lambda = \mu_V / \sigma_V = \sum_{i=1}^{t} c_i \mu_i \big/ \sqrt{\mathrm{Var}(\sum_{i=1}^{t} c_i P_i)} = \sum_{i=1}^{t} c_i \mu_i \big/ \sqrt{\sum_{i=1}^{t} c_i^2 \sigma_i^2 + 2 \sum_{i=1}^{t} \sum_{j \neq i} c_i c_j \sigma_{ij}}$, where σ_{ij} is the covariance of populations P_i and P_j. The c^+-probability for contrast V is $\mathrm{Pr}(V > 0)$. There is a strong relationship between c^+-probability and SMCV of a contrast variable V, which is derived from the following theorem [163;167;170].

Theorem 1. *Let U be a linear combination of random variables representing g groups with random values, namely $U = a_0 + \sum_{i=1}^{g} a_i G_i$, where a_0 is a known constant (usually $a_0 = 0$), a_i is the coefficient for the ith group, and G_i $(i = 1, \ldots, g)$ is a random variable with mean μ_i and variance σ_i^2 that represents the values in the ith group. The mean and variance of are μ_U and σ_U^2, respectively. Let W denote a standardized linear combination of U $\left(i.e., W = \frac{U - \mu_U}{\sigma_U}\right)$ and let $F(\cdot)$ and $\Phi(\cdot)$ be cumulative distribution functions of W and the standard normal distribution $N(0, 1)$, respectively. Define the parameter $\lambda = \frac{\mu_U}{\sigma_U}$ as the standardized mean of a linear combination (abbreviated as SMLC) and define the probability that U is greater than 0 as c^+-probability (i.e., c^+-probability $= \mathrm{Pr}(U > 0)$). Then, there exist the following relationships between SMLC and c^+-probability:*

1) For any distribution of U, c^+-probability $= 1 - F(-\lambda)$.

2) If U has normal distribution, c^+-probability $= \Phi(\lambda)$, where $\Phi(\cdot)$ is a cumulative distribution function of a standard normal distribution $N(0, 1)$.

3) If U has a unimodal distribution with non-zero finite variance,

$$
\begin{cases}
c^+\text{-probability} \geq 1 - \dfrac{4}{9\lambda^2}, & \text{for } \lambda \geq \sqrt{\dfrac{8}{3}} \\[2mm]
c^+\text{-probability} \geq \dfrac{4}{3} - \dfrac{4}{3\lambda^2}, & \text{for } 1 \leq \lambda \leq \sqrt{\dfrac{8}{3}} \\[2mm]
c^+\text{-probability} \leq \dfrac{4}{9\lambda^2}, & \text{for } \lambda \leq -\sqrt{\dfrac{8}{3}} \\[2mm]
c^+\text{-probability} \leq \dfrac{4}{3\lambda^2} - \dfrac{1}{3}, & \text{for } -1 \geq \lambda \geq -\sqrt{\dfrac{8}{3}}.
\end{cases}
$$

4) If U has a symmetric unimodal distribution with non-zero finite variance,

$$
\begin{cases}
c^+\text{-probability} \geq 1 - \dfrac{2}{9\lambda^2}, & \text{for } \lambda \geq \sqrt{\dfrac{8}{3}} \\[2mm]
c^+\text{-probability} \geq \dfrac{7}{6} - \dfrac{2}{3\lambda^2}, & \text{for } 1 \leq \lambda \leq \sqrt{\dfrac{8}{3}} \\[2mm]
c^+\text{-probability} \leq \dfrac{2}{9\lambda^2}, & \text{for } \lambda \leq -\sqrt{\dfrac{8}{3}} \\[2mm]
c^+\text{-probability} \leq \dfrac{2}{3\lambda^2} - \dfrac{1}{6}, & \text{for } -1 \geq \lambda \geq -\sqrt{\dfrac{8}{3}}.
\end{cases}
$$

Proof: It is trivial to show items 1) and 2) because $\Pr(U > 0) = \Pr\left(\frac{U - \mu_U}{\sigma_U} > -\frac{\mu_U}{\sigma_U}\right) = \Pr(W > -\lambda)$. The proof of items 3) and 4) relies on Vysochanskii-Petunin inequality [155]: for all $k > 0$, the following inequality holds for an arbitrary random variable X having a unimodal distribution and finite variance $\sigma^2 > 0$,

$$\begin{cases} \Pr(|X - \mu_X| \geq k\sigma) \leq \dfrac{4}{9k^2}, & \text{for } k \geq \sqrt{\dfrac{8}{3}} \\[2ex] \Pr(|X - \mu_X| \geq k\sigma) \leq \dfrac{4}{3k^2} - \dfrac{1}{3}, & \text{for } k \leq \sqrt{\dfrac{8}{3}}. \end{cases}$$

When $\lambda > 0$, $\Pr(U \leq 0) = \Pr(U - \mu_U \leq -\lambda\sigma_U) \leq \Pr(|U - \mu_U| \geq \lambda\sigma_U)$. In the situation where U has a unimodal distribution with finite variance $\sigma_U^2 > 0$, applying Vysochanskii-Petunin inequality with $k = \lambda$ to variable U, we get

$$\begin{cases} \Pr(|U - \mu_U| \geq \lambda\sigma_U) \leq \dfrac{4}{9\lambda^2}, & \text{for } \lambda \geq \sqrt{\dfrac{8}{3}} \\[2ex] \Pr(|U - \mu_U| \geq \lambda\sigma_U) \leq \dfrac{4}{3\lambda^2} - \dfrac{1}{3}, & \text{for } 1 \leq \lambda \leq \sqrt{\dfrac{8}{3}}. \end{cases}$$

Considering c^+-probability $= \Pr(U > 0) = 1 - \Pr(U \leq 0)$, we then have

$$\begin{cases} c^+\text{-probability} \geq 1 - \dfrac{4}{9\lambda^2}, & \text{for } \lambda \geq \sqrt{\dfrac{8}{3}} \\[2ex] c^+\text{-probability} \geq \dfrac{4}{3} - \dfrac{4}{3\lambda^2}, & \text{for } 1 \leq \lambda \leq \sqrt{\dfrac{8}{3}}. \end{cases}$$

Similarly, when $\lambda < 0$, $\Pr(U > 0) = \Pr(U - \mu_U > -\lambda\sigma_U) \leq \Pr(|U - \mu_U| \geq (-\lambda) \cdot \sigma_U)$. Applying Vysochanskii-Petunin inequality with $k = -\lambda$ to variable V, we get

$$\begin{cases} c^+\text{-probability} \leq \dfrac{4}{9\lambda^2}, & \text{for } \lambda \leq -\sqrt{\dfrac{8}{3}} \\[2ex] c^+\text{-probability} \leq \dfrac{4}{3\lambda^2} - \dfrac{1}{3}, & \text{for } -1 \geq \lambda \geq -\sqrt{\dfrac{8}{3}}. \end{cases}$$

Thus we prove item 3); similarly, item 4) can be proved.

Set $a_0 = 0$, $a_i = c_i$, and $g = t$. Then U becomes a contrast variable V, and SMLC becomes SMCV. Thus the relationships between SMLC and c^+-probability in the preceding theorem become the relationships between SMCV and c^+-probability, which are shown in Figure 7.2.

7.3 A Classifying Rule for Interpreting Strength of Group Comparisons

Clear and consistent interpretations for the strength of group comparisons are an important and urgent need, which is reflected in the comment by Rosenthal, Rosnow,

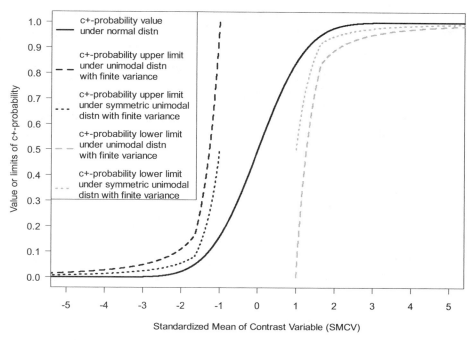

Figure 7.2   The relationships between SMCV and c^+-probability.

and Rubin [126], "Despite the growing awareness of the importance of estimating sizes of effects along with obtaining levels of significance, problems of interpretation remain," as well as in the comment by Huberty [72], "The interpretation of the index value magnitude is, perhaps, the biggest limitation of the use of effect size." Based on contrast variable, SMCV, and c^+-probability, Zhang [167] provides a clear and consistent interpretation to the strength of a comparison.

With a contrast variable, we can assess the strength of a contrast from two aspects: one is based on the integration of both mean and variability of a contrast variable, represented by SMCV, and the other is based on distributions in multiple conditions, represented by c^+-probability. More importantly, because of the relationship between SMCV and c^+-probability, we can classify the strength of contrast based on SMCV, which simultaneously contains information from c^+-probability. Based on SMCV, some key values of interest are 0.25, 0.5, 0.75, 1, 2, 3, and 5, which means that the average value of a contrast variable is one quarter, one half (or two quarters), three quarters of and one time, two times, three times, and five times the standard deviation of the contrast variable. Under normality, their corresponding values of c^+-probability are 0.60, 0.69, 0.77, 0.84, 0.97725, 0.99865, and 0.9999997, respectively. Based on c^+-probability, the values of interest for SMCV are 1.28, 1.645, 3, and 5. This is because SMCV = 1.28 and 1.645 corresponds to c^+-probability = 0.90 and 0.95, respectively, under normality, and SMCV = 3 and 5 indicates that the values of corresponding c^+-probability are at least 0.975 and 0.99, respectively, when

Table 7.1. Key values of SMCV and the value/limit of their corresponding c^+-probability

	A. Positive SMCV and c^+-Probability				B. Negative SMCV and c^+-Probability		
	c^+-Probability				c^+-Probability		
	Value	Lower Limit			Value	Upper Limit	
SMCV	Normal	Symmetric Unimodal	Unimodal	SMCV	Normal	Symmetric Unimodal	Unimodal
0	0.50			−0	0.50		
0.25	0.60			−0.25	0.40		
0.50	0.69			−0.50	0.31		
0.75	0.77			−0.75	0.23		
1	0.84	0.5	0	−1	0.16	0.5	1
1.28	0.90	0.76	0.52	−1.28	0.10	0.24	0.48
1.645	0.95	0.918	0.836	−1.645	0.05	0.082	0.164
2	0.97725	0.944	0.89	−2	0.02275	0.056	0.112
3	0.99865	0.975	0.95	−3	0.00135	0.025	0.05
5		0.99	0.98	−5		0.01	0.02

the contrast variable has a symmetric unimodal distribution (Table 7.1) and at least 0.95 and 0.98, respectively, when the contrast variable has a unimodal distribution. Note that 1.645 equals approximately $\sqrt{8/3}$, an important value in the relationship between SMCV and c^+-probability for non-normal distributions (Figure 7.2). Therefore, based on both SMCV and c^+-probability, the key values of interest for SMCV are 0, 0.25, 0.5, 0.75, 1, 1.28, 1.645, 2, 3, and 5.

Based on the key values of SMCV and their corresponding values of c^+-probability (Table 7.1), it is reasonable to construct the SMCV-based criteria listed in Table 7.2 for assessing the strength of a comparison. For example, when the mean of a contrast variable is zero (i.e., SMCV = 0), c^+-probability is 0.5. When SMCV is between 0 and 0.25, c^+-probability is between 0.50 and 0.60, which is slightly above 0.50, and thus the strength of the contrast is extremely weak. Similarly, other categories in the second column of Table 7.2 are constructed based on the values of SMCV and c^+-probability. There are 10 categories of effect types in the second column of Table 7.2. In social sciences, researchers may prefer fewer categories, similarly to Cohen's or McLean's criterion [33;104]. In such a case, one may adopt the effect types listed in the first column of Table 7.2. The criteria in Table 7.2 can be applied to any linear combination of random variables in which the interest is how far the linear combination is away from zero. However, they work in the situation in which the population value of SMCV is known. In practice, the population value of SMCV is unknown; thus we need to estimate SMCV and c^+-probability.

Table 7.2. SMCV-based criteria for classifying the strength of a contrast

Effect Types	Effect Subtypes	Thresholds for Positive SMCV	Thresholds for Negative SMCV
Extra large	Extremely strong	$\lambda \geq 5$	$\lambda \leq -5$
	Very strong	$5 > \lambda \geq 3$	$-5 < \lambda \leq -3$
	Strong	$3 > \lambda \geq 2$	$-3 < \lambda \leq -2$
	Fairly strong	$2 > \lambda \geq 1.645$	$-2 < \lambda \leq -1.645$
Large	Moderate	$1.645 > \lambda \geq 1.28$	$-1.645 < \lambda \leq -1.28$
	Fairly moderate	$1.28 > \lambda \geq 1$	$-1.28 < \lambda \leq -1$
Medium large	Fairly weak	$1 > \lambda \geq 0.75$	$-1 < \lambda \leq -0.75$
	Weak	$0.75 > \lambda > 0.5$	$-0.75 < \lambda < -0.5$
Medium	Very weak	$0.5 \geq \lambda > 0.25$	$-0.5 \leq \lambda < -0.25$
Small	Extremely weak	$0.25 \geq \lambda > 0$	$-0.25 \leq \lambda < 0$
	No effect	$\lambda = 0$	

Note: λ denotes standardized mean of contrast variable (SMCV).

7.4 A Theorem to Facilitate the Estimation and Inference of SMCV

To facilitate the estimation and inference of SMCV, I first provide the following theorem for a general linear combination of random variables.

Theorem 2. *Consider the situation in which g groups are all independently and normally distributed, namely $G_i \sim N(\mu_i, \sigma_i^2)$, $i = 1, \ldots, g$ independently. Random samples are independently obtained from the g groups, $Y_i = (Y_{i1}, Y_{i2}, \ldots, Y_{in_i})$, $i = 1, \ldots, g$. The sample size, mean, and variance of the sample from group G_i are n_i, $\bar{Y}_i$, and s_i^2, respectively. Let μ and λ_U respectively be the mean and SMLC of a linear combination $U = a_0 + \sum_{i=1}^{g} a_i G_i$. Then, the following properties hold for a parameter $\lambda = \frac{1}{\sqrt{m}} \lambda_U = \frac{1}{\sqrt{m}} \left(a_0 + \sum_{i=1}^{g} a_i \mu_i \right) \Big/ \sqrt{\sum_{i=1}^{g} a_i^2 \sigma_i^2}$ (where m is a constant).*
1) When the variances of the g groups are not equal:
a) The maximum likelihood estimate (MLE) of λ is

$$\hat{\lambda}_{\mathrm{MLE}} = \frac{1}{\sqrt{m}} \frac{a_0 + \sum\limits_{i=1}^{g} a_i \bar{Y}_i}{\sqrt{\sum\limits_{i=1}^{g} \dfrac{n_i - 1}{n_i} a_i^2 s_i^2}}, \tag{T2.1}$$

and the method-of-moment estimate (MM) of λ is

$$\hat{\lambda}_{\mathrm{MM}} = \frac{1}{\sqrt{m}} \frac{a_0 + \sum\limits_{i=1}^{g} a_i \bar{Y}_i}{\sqrt{\sum\limits_{i=1}^{g} a_i^2 s_i^2}}, \tag{T2.2}$$

b) $\hat{\lambda}_{\text{MLE}}$ *is asymptotically distributed with a normal distribution, namely*

$$\hat{\lambda}_{\text{MLE}} \xrightarrow{D} N\left(\lambda, \frac{1}{m}\left(\frac{\sum\limits_{i=1}^{g}\frac{a_i^2\sigma_i^2}{n_i}}{\sum\limits_{i=1}^{g}a_i^2\sigma_i^2} + \frac{1}{2}\frac{\sum\limits_{i=1}^{g}\frac{a_i^4\sigma_i^4}{n_i}}{\left(\sum\limits_{i=1}^{g}a_i^2\sigma_i^2\right)^3}\cdot\left(a_0+\sum\limits_{i=1}^{g}a_i\mu_i\right)^2\right)\right)$$

(T2.3)

c) *When the sample size equals r in each group, namely* $n_1 = \cdots = n_g = r$,

$$\frac{a_0+\sum\limits_{i=1}^{g}a_i\bar{Y}_i}{\sqrt{\frac{1}{r}\sum\limits_{i=1}^{g}a_i^2 s_i^2}} \sim t\left(v, \lambda\sqrt{mr}\right) \text{ approximately, } v = (r-1)\frac{\left(\sum\limits_{i=1}^{g}a_i^2 s_i^2\right)^2}{\sum\limits_{i=1}^{g}a_i^4 s_i^4}$$

(T2.4)

One approximately unbiased estimate of λ *is*

$$\hat{\lambda}_{\text{AUE}} = \sqrt{\frac{2}{v}}\frac{\Gamma\left(\frac{v}{2}\right)}{\Gamma\left(\frac{v-1}{2}\right)}\frac{1}{\sqrt{m}}\frac{a_0+\sum\limits_{i=1}^{g}a_i\bar{Y}_i}{\sqrt{\sum\limits_{i=1}^{g}a_i^2 s_i^2}}$$

(T2.5)

2) *When the variances of the g groups are equal:*
   a) *The uniformly minimal variance unbiased estimate (UMVUE), MLE, and MM estimates of* λ *are, respectively,*

$$\hat{\lambda}_{\text{UMVUE}} = \frac{\sqrt{K}}{\sqrt{n-g}}\frac{1}{\sqrt{m}}\frac{a_0+\sum\limits_{i=1}^{g}a_i\bar{Y}_i}{\sqrt{\text{MSE}\cdot\sum\limits_{i=1}^{g}a_i^2}}$$

(T2.6)

$$\hat{\lambda}_{\text{MLE}} = \sqrt{\frac{n}{n-g}}\frac{1}{\sqrt{m}}\frac{a_0+\sum\limits_{i=1}^{g}a_i\bar{Y}_i}{\sqrt{\text{MSE}\cdot\sum\limits_{i=1}^{g}a_i^2}}$$

(T2.7)

$$\hat{\lambda}_{\text{MM}} = \frac{1}{\sqrt{m}}\frac{a_0+\sum\limits_{i=1}^{g}a_i\bar{Y}_i}{\sqrt{\text{MSE}\cdot\sum\limits_{i=1}^{g}a_i^2}}$$

(T2.8)

where $K = 2 \cdot \left(\dfrac{\Gamma\left(\frac{n-g}{2}\right)}{\Gamma\left(\frac{n-g-1}{2}\right)} \right)^2$, $n = \sum\limits_{i=1}^{g} n_i$ and $\text{MSE} = \frac{1}{n-g} \sum_{i=1}^{g} (n_i - 1)s_i^2$.

b) We have the following noncentral t-distribution

$$\frac{a_0 + \sum\limits_{i=1}^{g} a_i \bar{Y}_i}{\sqrt{\text{MSE} \cdot \sum\limits_{i=1}^{g} \dfrac{a_i^2}{n_i}}} \sim \text{noncentral } t\left(n - g, \lambda \sqrt{m \sum_{i=1}^{g} a_i^2} \Big/ \sqrt{\sum_{i=1}^{g} \frac{a_i^2}{n_i}} \right) \quad \text{(T2.9)}$$

Proof: Let $Y_i = (Y_{i1}, Y_{i2}, \ldots, Y_{in_i})$, $i = 1, \ldots, g$ be the random sample from group G_i; $f_i(d_{i1}; \theta_i)$ be the probability density function of Y_{i1}; and $\theta_i = \left(\mu_i, \sigma_i^2\right)^T$. As shown in many classical textbooks, the MLE of μ_i and σ_i^2 are $\hat{Y}_i$ and $\frac{n_i-1}{n_i}s_i^2$, respectively; that is, the MLE of θ_i is $\hat{\theta}_i = \left(\bar{Y}_i, \frac{n_i-1}{n_i}s_i^2 \right)^T$. By the invariance property of MLE, the MLE of λ is $\hat{\lambda}_{\text{MLE}} = \frac{1}{\sqrt{m}} \left(a_0 + \sum_{i=1}^{g} a_i \bar{Y}_i \right) \Big/ \sqrt{\sum_{i=1}^{g} \frac{n_i-1}{n_i} a_i^2 s_i^2}$. Thus Formula T2.1 is proved. Formula T2.2 directly comes from the definition of MM estimate.

By the asymptotical normality and efficiency of MLE from the identically independently distributed (IID) sample Y_i, we have $\sqrt{n_i}(\hat{\theta}_i - \theta_i) \xrightarrow{D} N\left(0, I^{-1}(\theta_i)\right)$ as $n_i \longrightarrow \infty$, where $I(\theta_i) = \text{E}\left(-\frac{\partial^2}{\partial \theta_i^2} \log(f_i(Y_{i1}; \theta_i))\right)$. It is trivial to show that $I(\theta_i) = \begin{pmatrix} \frac{1}{\sigma_i^2} & 0 \\ 0 & \frac{1}{2\sigma_i^4} \end{pmatrix}$. Therefore, as $n_i \longrightarrow \infty$,

$$\sqrt{n_i}\left(\begin{pmatrix} \bar{Y}_i \\ \dfrac{n_i - 1}{n_i}s_i^2 \end{pmatrix} - \begin{pmatrix} \mu_i \\ \sigma_i^2 \end{pmatrix} \right) \xrightarrow{D} \left(\begin{pmatrix} 0 \\ 0 \end{pmatrix}, \begin{pmatrix} \sigma_i^2 & 0 \\ 0 & 2\sigma_i^4 \end{pmatrix} \right);$$

subsequently,

$$\left(\begin{pmatrix} \bar{Y}_1 \\ \dfrac{n_1 - 1}{n_1}s_1^2 \\ \vdots \\ \bar{Y}_g \\ \dfrac{n_g - 1}{n_g}s_g^2 \end{pmatrix} - \begin{pmatrix} \mu_1 \\ \sigma_1^2 \\ \vdots \\ \mu_g \\ \sigma_g^2 \end{pmatrix} \right) \xrightarrow{D} N\left(\begin{pmatrix} 0 \\ 0 \\ \vdots \\ 0 \\ 0 \end{pmatrix}, \begin{pmatrix} \sigma_1^2/n_1 & 0 & \cdots & 0 & 0 \\ 0 & 2\sigma_1^4/n_1 & \cdots & 0 & 0 \\ \vdots & \vdots & \ddots & \vdots & \vdots \\ 0 & 0 & \cdots & \sigma_g^2/n_g & 0 \\ 0 & 0 & \cdots & 0 & 2\sigma_g^4/n_g \end{pmatrix} \right).$$

By Delta method,

$$(\hat{\lambda}_{\text{MLE}} - \lambda) \xrightarrow{D} N\left(0, \frac{\partial \lambda}{\partial \theta^T} \begin{pmatrix} \sigma_1^2/n_1 & 0 & \cdots & 0 & 0 \\ 0 & 2\sigma_1^4/n_1 & \cdots & 0 & 0 \\ \vdots & \vdots & \ddots & \vdots & \vdots \\ 0 & 0 & \cdots & \sigma_g^2/n_g & 0 \\ 0 & 0 & \cdots & 0 & 2\sigma_g^4/n_g \end{pmatrix} \frac{\partial \lambda}{\partial \theta} \right)$$

where $\theta = \left(\mu_1, \sigma_1^2, \ldots, \mu_g, \sigma_g^2\right)^T$ and

$$\frac{\partial \lambda}{\partial \theta^T} = \left(\frac{\partial \lambda}{\partial \mu_1}, \frac{\partial \lambda}{\partial \sigma_1^2}, \ldots, \frac{\partial \lambda}{\partial \mu_g}, \frac{\partial \lambda}{\partial \sigma_g^2}\right)$$

$$= \frac{1}{\sqrt{m}} \left(\frac{a_1}{\sqrt{\displaystyle\sum_{i=1}^{g} a_i^2 \sigma_i^2}}, -\frac{1}{2} \frac{a_1^2 \left(a_0 + \displaystyle\sum_{i=1}^{g} a_i \mu_i\right)}{\left(\displaystyle\sum_{i=1}^{g} a_i^2 \sigma_i^2\right)^{\frac{3}{2}}}, \ldots, \right.$$

$$\left. \frac{a_g}{\sqrt{\displaystyle\sum_{i=1}^{g} a_i^2 \sigma_i^2}}, -\frac{1}{2} \frac{a_g^2 \left(a_0 + \displaystyle\sum_{i=1}^{g} a_i \mu_i\right)}{\left(\displaystyle\sum_{i=1}^{g} a_i^2 \sigma_i^2\right)^{\frac{3}{2}}} \right)$$

Therefore,

$$\hat{\lambda}_{\text{MLE}} \xrightarrow{D} N\left(\lambda, \sigma_{\hat{\lambda}}^2\right)$$

where

$$\sigma_{\hat{\lambda}}^2 = \frac{1}{m} \left(\frac{\displaystyle\sum_{i=1}^{g} \frac{a_i^2 \sigma_i^2}{n_i}}{\displaystyle\sum_{i=1}^{g} a_i^2 \sigma_i^2} + \frac{1}{2} \frac{\displaystyle\sum_{i=1}^{g} \frac{a_i^4 \sigma_i^4}{n_i}}{\left(\displaystyle\sum_{i=1}^{g} a_i^2 \sigma_i^2\right)^3} \cdot \left(a_0 + \displaystyle\sum_{i=1}^{g} a_i \mu_i\right)^2 \right)$$

and (T2.3) is proved similarly as in [162].

The proof of the remaining parts in Theorem 2 requires the use of the well-known results below.

- Result 1: if X_i's are independently distributed with normal distributions, that, $X_i \sim N(\mu_i, \sigma_i^2)$, then

$$a_0 + \sum_{i=1}^{n} c_i X_i \sim N\left(a_0 + \sum_{i=1}^{n} c_i \mu_i, \sum_{i=1}^{n} c_i^2 \sigma_i^2\right).$$

- Result 2 (Satterthwaite approximation [110;129]): for n sample variances s_i^2 ($i = 1, \ldots, n$), each having ν_i degrees of freedom, the linear combination of sample variances approximately has a χ^2-distribution. That is:

$$\frac{\nu \displaystyle\sum_{i=1}^{n} c_i s_i^2}{\displaystyle\sum_{i=1}^{n} c_i \sigma_i^2} \sim \chi^2(\nu) \text{ approximately, where } \nu = \frac{\left(\displaystyle\sum_{i=1}^{n} c_i s_i^2\right)^2}{\displaystyle\sum_{i=1}^{n} \frac{\left(c_i s_i^2\right)^2}{\nu_i}}.$$

- Result 3: if two random variables X and U are independently distributed with $X \sim N(\mu, 1)$ and $U \sim \chi^2(p)$, then the ratio $T = X/\sqrt{U/p}$ has a noncentral t-distribution with p degrees of freedom and noncentrality parameter μ, namely $T \sim$ noncentral $t(p, \mu)$.

- Result 4: the mean and variance of noncentral $t(p, \mu)$ are $(p/2)^{\frac{1}{2}} \dfrac{\Gamma((p-1)/2)}{\Gamma(p/2)} \mu$

  and $\dfrac{p}{p-2} + \left(\dfrac{p}{p-2} - \dfrac{p}{2} \left(\dfrac{\Gamma(\frac{p-1}{2})}{\Gamma(\frac{p}{2})} \right)^2 \right) \mu^2$, respectively.

  Because of $\bar{Y}_i \sim N(\mu_i, \frac{\sigma_i^2}{n_i})$ and Result 1, we have

$$a_0 + \sum_{i=1}^{g} a_i \bar{Y}_i \sim N\left(\mu, \sum_{i=1}^{g} \frac{a_i^2 \sigma_i^2}{n_i} \right);$$

subsequently

$$\left(a_0 + \sum_{i=1}^{g} a_i \bar{Y}_i \right) \Big/ \sqrt{ \sum_{i=1}^{g} \frac{a_i^2 \sigma_i^2}{n_i} } \sim N\left(\mu \Big/ \sqrt{ \sum_{i=1}^{g} \frac{a_i^2 \sigma_i^2}{n_i} }, 1 \right).$$

Applying Result 2 with $c_i = \dfrac{a_i^2}{n_i}$, we have

$$\left(\nu \sum_{i=1}^{g} \frac{a_i^2}{n_i} s_i^2 \Big/ \sum_{i=1}^{g} \frac{a_i^2}{n_i} \sigma_i^2 \right) \sim \chi^2(\nu)$$

approximately, where

$$\nu = \frac{ \left(\sum_{i=1}^{g} \frac{a_i^2}{n_i} s_i^2 \right)^2 }{ \sum_{i=1}^{g} \frac{ \left(\frac{a_i^2}{n_i} s_i^2 \right)^2 }{ \nu_i } } = \frac{ \left(\sum_{i=1}^{g} \frac{a_i^2 s_i^2}{n_i} \right)^2 }{ \sum_{i=1}^{g} \frac{a_i^4 s_i^4}{n_i^2 (n_i - 1)} }.$$

Then using Result 3, we have

$$\frac{ \left(a_0 + \sum_{i=1}^{g} a_i \bar{Y}_i \right) \Big/ \sqrt{ \sum_{i=1}^{g} \frac{a_i^2 \sigma_i^2}{n_i} } }{ \sqrt{ \sum_{i=1}^{g} \frac{a_i^2 s_i^2}{n_i} \Big/ \sum_{i=1}^{g} \frac{a_i^2 \sigma_i^2}{n_i} } } \sim t\left(\nu, \mu \Big/ \sqrt{ \sum_{i=1}^{g} \frac{a_i^2 \sigma_i^2}{n_i} } \right)$$

approximately where

$$\nu = \frac{ \left(\sum_{i=1}^{g} \frac{a_i^2 s_i^2}{n_i} \right)^2 }{ \sum_{i=1}^{g} \frac{a_i^4 s_i^4}{n_i^2 (n_i - 1)} },$$

namely

$$\frac{a_0 + \sum_{i=1}^{g} a_i \bar{Y}_i}{\sqrt{\sum_{i=1}^{g} \frac{a_i^2 s_i^2}{n_i}}} \sim t\left(\nu, \mu \bigg/ \sqrt{\sum_{i=1}^{g} \frac{a_i^2 \sigma_i^2}{n_i}}\right)$$

approximately. If $n_1 = \cdots = n_g = r$, then

$$T = \frac{a_0 + \sum_{i=1}^{g} a_i \bar{Y}_i}{\sqrt{\frac{1}{r} \sum_{i=1}^{g} a_i^2 s_i^2}} \sim t\left(\nu, \lambda \cdot \sqrt{m \sum_{i=1}^{g} a_i^2 \sigma_i^2} \bigg/ \sqrt{\frac{1}{r} \sum_{i=1}^{g} a_i^2 \sigma_i^2}\right) = t(\nu, \sqrt{mr}\lambda)$$

approximately where $\nu = (r-1) \times \left(\sum_{i=1}^{g} a_i^2 s_i^2\right)^2 / \sum_{i=1}^{g} a_i^4 s_i^4$. Thus Formula T2.4 is proved.

By Result 4,

$$E(T) = E\left(\frac{a_0 + \sum_{i=1}^{g} a_i \bar{Y}_i}{\sqrt{\sum_{i=1}^{g} \frac{a_i^2 s_i^2}{n_i}}}\right) \approx E\left(t(\nu, \sqrt{mr}\lambda)\right) = \left(\tfrac{\nu}{2}\right)^{\frac{1}{2}} \frac{\Gamma\left(\frac{\nu-1}{2}\right)}{\Gamma\left(\frac{\nu}{2}\right)} \sqrt{mr}\lambda.$$

If set

$$\hat{\lambda} = \sqrt{\frac{2}{r\nu}} \frac{\Gamma\left(\frac{\nu}{2}\right)}{\Gamma\left(\frac{\nu-1}{2}\right)} \frac{1}{\sqrt{m}} \frac{a_0 + \sum_{i=1}^{g} a_i \bar{Y}_i}{\sqrt{\sum_{i=1}^{g} \frac{a_i^2 s_i^2}{n_i}}} = \sqrt{\frac{2}{\nu}} \frac{\Gamma\left(\frac{\nu}{2}\right)}{\Gamma\left(\frac{\nu-1}{2}\right)} \frac{1}{\sqrt{m}} \frac{a_0 + \sum_{i=1}^{g} a_i \bar{Y}_i}{\sqrt{\sum_{i=1}^{g} a_i^2 s_i^2}},$$

then $E(\hat{\lambda}) \approx \lambda$. This is why we have the estimate in Formula T2.5.

When the group G_i's have equal variance $\left(\text{i.e., } \sigma_1^2 = \sigma_2^2 = \cdots = \sigma_g^2 = \sigma_e^2\right)$, G_i's are independently and normally distributed with $G_i \sim N(\mu_i, \sigma_e^2)$, which leads to the common model $G_{ij} = \mu_i + \varepsilon_{ij}$ and $\varepsilon_{ij} \sim N(0, \sigma_e^2)$ in one-way ANOVA. It is well-known that the following properties hold in the situation with equal variance:

1) $\bar{Y}_1, \ldots, \bar{Y}_g, s_1^2, \ldots, s_g^2$ are all independent with each other.

2) $\frac{(n-g)\cdot \text{MSE}}{\sigma_e^2} \sim \chi^2(n-g)$, subsequently $E\left(1 / \sqrt{\frac{(n-g)\cdot \text{MSE}}{\sigma_e^2}}\right) = \frac{1}{\sqrt{K}}$, where

$$K = 2 \cdot \left(\frac{\Gamma\left(\frac{n-g}{2}\right)}{\Gamma\left(\frac{n-g-1}{2}\right)}\right)^2, \quad n = \sum_{i=1}^{g} n_i,$$

and $\text{MSE} = \frac{1}{n-g} \sum_{i=1}^{g} (n_i - 1)s_i^2$.

3) $\left(\bar{Y}_1, \ldots, \bar{Y}_g, s_1^2, \ldots, s_g^2\right)$ is a complete sufficient statistic of $(\mu_1, \ldots, \mu_g, \sigma_e^2)$.

Based on properties 1 and 2,

$$
E\left(\frac{1}{\sqrt{m}}\frac{\sqrt{K}}{\sqrt{n-g}}\frac{a_0+\sum_{i=1}^{g}a_i\bar{Y}_i}{\sqrt{\mathrm{MSE}\sum_{i=1}^{g}a_i^2}}\right)=\frac{1}{\sqrt{m\sum_{i=1}^{g}a_i^2}}\cdot E\left(a_0+\sum_{i=1}^{g}a_i\bar{Y}_i\right).
$$

$$
E\left(\frac{\sqrt{K}}{\sqrt{\mathrm{MSE}\cdot(n-g)}}\right)=\frac{1}{\sqrt{m}}\frac{a_0+\sum_{i=1}^{g}a_i\mu_i}{\sigma_e\sqrt{\sum_{i=1}^{g}a_i^2}}=\lambda
$$

where $n_1,\ldots,n_g\geq 2$. Set $\hat{\lambda}=\frac{1}{\sqrt{m}}\frac{\sqrt{K}}{\sqrt{n-g}}\left(a_0+\sum_{i=1}^{g}a_i\bar{Y}_i\right)/\sqrt{\mathrm{MSE}\sum_{i=1}^{g}a_i^2}$. Then $\hat{\lambda}$ is an unbiased estimate of λ and is a function of complete sufficient statistic $(\bar{Y}_1,\ldots,\bar{Y}_g,s_1^2,\ldots,s_g^2)$; thus $\hat{\lambda}$ is the UMVUE of pooled SSMD λ. Therefore, T2.6 is proved. The proof of T2.7 and T2.8 is trivial with the consideration that the MLE of σ_e^2 is $\frac{n-g}{n}\mathrm{MSE}$.

Using $\bar{Y}_i\sim N\left(\mu_i,\frac{\sigma_e^2}{n_i}\right)$ and Result 1, $a_0+\sum_{i=1}^{g}a_i\bar{Y}_i\sim N\left(\mu,\sigma_e^2\sum_{i=1}^{g}\frac{a_i^2}{n_i}\right)$; subsequently,

$$
\left(a_0+\sum_{i=1}^{g}a_i\bar{Y}\right)\bigg/\sqrt{\sigma_e^2\sum_{i=1}^{g}\frac{a_i^2}{n_i}}\sim N\left(\mu\bigg/\sqrt{\sigma_e^2\sum_{i=1}^{g}\frac{a_i^2}{n_i}},1\right).
$$

Using the above property 2 and result 3, we have

$$
T=a_0+\sum_{i=1}^{g}a_i\bar{Y}_i\bigg/\sqrt{\mathrm{MSE}\cdot\sum_{i=1}^{g}\frac{a_i^2}{n_i}}\sim \text{noncentral }t\left(n-g,\mu\bigg/\sqrt{\sigma_e^2\sum_{i=1}^{g}\frac{a_i^2}{n_i}}\right),
$$

namely $T\sim$ noncentral $t\left(n-g,\lambda\sqrt{m\sigma_e^2\sum_{i=1}^{g}a_i^2}\big/\sqrt{\sigma_e^2\sum_{i=1}^{g}\frac{a_i^2}{n_i}}\right)$. Therefore Formula T2.9 is proved.

The following results are for the estimation and inference of the mean μ of the linear combination of random variables.

Whether the variances of the groups are equal or not, the MLE of μ_i is $\bar{Y}_i$. By the invariance property of MLE, the MLE of μ is $\hat{\mu}_{\mathrm{MLE}}=a_0+\sum_{i=1}^{g}a_i\bar{Y}_i$. By the definition of MM estimate, $\hat{\mu}_{\mathrm{MM}}=a_0+\sum_{i=1}^{g}a_i\bar{Y}_i$. Considering (i) $(\bar{Y}_1,\bar{Y}_2,\ldots,\bar{Y}_g)$ is a complete sufficient statistic of $(\mu_1,\mu_2,\ldots,\mu_g)$, (ii) $a_0+\sum_{i=1}^{g}a_i\bar{Y}_i$ is a function of $(\bar{Y}_1,\bar{Y}_2,\ldots,\bar{Y}_g)$, and (iii) $E\left(a_0+\sum_{i=1}^{g}a_i\bar{Y}_i\right)=\mu$, we have $\hat{\mu}_{\mathrm{UMVUE}}=a_0+\sum_{i=1}^{g}a_i\bar{Y}_i$.

When the variances of the g groups are not equal, $\bar{Y}_i\sim N\left(\mu_i,\sigma_i^2/n_i\right)$. By result 1,

$$
a_0+\sum_{i=1}^{g}a_i\bar{Y}_i\sim N\left(\mu,\sum_{i=1}^{g}\frac{a_i^2\sigma_i^2}{n_i}\right).
$$

Subsequently,

$$\left(a_0 + \sum_{i=1}^{g} a_i \bar{Y}_i - \mu\right) \Big/ \sqrt{\sum_{i=1}^{g} \frac{a_i^2 \sigma_i^2}{n_i}} \sim N(0, 1).$$

It is well-known that if two random variables X and U are independently distributed with $X \sim N(0, 1)$ and $U \sim \chi^2(p)$, then the ratio $T = X/\sqrt{U/p}$ has a central t-distribution with p degrees of freedom, namely $T \sim t(p)$. Therefore, using the above distribution and result 2, we have

$$\frac{\left(a_0 + \sum_{i=1}^{g} a_i \bar{Y}_i - \mu\right) \Big/ \sqrt{\sum_{i=1}^{g} \frac{a_i^2 \sigma_i^2}{n_i}}}{\sqrt{\left(\nu \sum_{i=1}^{g} \frac{a_i^2 s_i^2}{n_i} \Big/ \sum_{i=1}^{g} \frac{a_i^2 \sigma_i^2}{n_i}\right) / \nu}} \sim t(\nu)$$

approximately where

$$\nu = \frac{\left(\sum_{i=1}^{g} \frac{a_i^2 s_i^2}{n_i}\right)^2}{\sum_{i=1}^{g} \frac{\left(a_i^2 s_i^2 / n_i\right)^2}{n_i - 1}}.$$

That is, $\left(a_0 + \sum_{i=1}^{g} a_i \bar{Y}_i - \mu\right) \Big/ \sqrt{\sum_{i=1}^{g} \left(a_i^2 s_i^2 / n_i\right)} \sim t(\nu)$ approximately. Under the null hypothesis $H_0 : \mu = 0$, $\left(a_0 + \sum_{i=1}^{g} a_i \bar{Y}_i\right) \Big/ \sqrt{\sum_{i=1}^{g} \left(a_i^2 s_i^2\right)/(n_i)} \sim t(\nu)$, which leads to the well-known Welch t-test in the case of $a_0 = 0$, $a_1 = 1$, $a_2 = -1$.

When the variances of the groups are equal, $\bar{Y}_i \sim N\left(\mu_i, \sigma_e^2 / n_i\right)$. By result 1, $a_0 + \sum_{i=1}^{g} a_i \bar{Y}_i \sim N\left(\mu, \sum_{i=1}^{g} \left(a_i^2 \sigma_e^2 / n_i\right)\right)$. Subsequently,

$$\left(a_0 + \sum_{i=1}^{g} a_i \bar{Y}_i - \mu\right) \Big/ \sqrt{\sum_{i=1}^{g} \frac{a_i^2 \sigma_e^2}{n_i}} \sim N(0, 1).$$

We know that $(n - g) \cdot \text{MSE}/\sigma_e^2 \sim \chi^2(n - g)$, where $\text{MSE} = \frac{1}{n-g} \sum_{i=1}^{g} (n_i - 1) s_i^2$.

Using result 3, we have

$$\frac{\left(a_0 + \sum_{i=1}^{g} a_i \bar{Y}_i - \mu\right) \Big/ \sqrt{\sum_{i=1}^{g} \frac{a_i^2 \sigma_e^2}{n_i}}}{\sqrt{\frac{(n - g) \cdot \text{MSE}}{\sigma_e^2} / (n - g)}} \sim t(n - g).$$

That is,

$$\frac{a_0 + \sum_{i=1}^{g} a_i \bar{Y}_i - \mu}{\sqrt{\mathrm{MSE} \cdot \sum_{i=1}^{g} \frac{a_i^2}{n_i}}} = \frac{a_0 + \sum_{i=1}^{g} a_i \bar{Y}_i - \mu}{\sqrt{\frac{1}{n-g} \sum_{i=1}^{g} (n_i - 1)s_i^2 \cdot \sum_{i=1}^{g} \frac{a_i^2}{n_i}}} \sim t(n - g)$$

In summary, we have the following results for the estimation and inference of the mean μ of a linear combination of random variables in a situation in which the g groups are independently normally distributed:

1) Whether the variances of the g groups are equal or not, the UMVUE, MLE, and MM estimates of μ are all

$$\hat{\mu}_{\mathrm{UMVUE}} = \hat{\mu}_{\mathrm{MLE}} = \hat{\mu}_{\mathrm{MM}} = a_0 + \sum_{i=1}^{g} a_i \bar{Y}_i \qquad \text{(T2.10)}$$

2) When the variances of the groups are not equal,

$$\left(a_0 + \sum_{i=1}^{g} a_i \bar{Y}_i - \mu \right) \Big/ \sqrt{\sum_{i=1}^{g} \frac{a_i^2 s_i^2}{n_i}} \sim t(\nu) \ \text{ approximately.} \qquad \text{(T2.11)}$$

Thus the $1 - \alpha$ confidence interval of μ is approximately

$$a_0 + \sum_{i=1}^{g} a_i \bar{Y}_i \pm t_{\nu, 1-\alpha/2} \sqrt{\sum_{i=1}^{g} \frac{a_i^2 s_i^2}{n_i}}. \qquad \text{(T2.12)}$$

3) When the variances of the groups are equal,

$$\frac{a_0 + \sum_{i=1}^{g} a_i \bar{Y}_i - \mu}{\sqrt{\frac{1}{n-g} \sum_{i=1}^{g} (n_i - 1)s_i^2 \cdot \sum_{i=1}^{g} \frac{a_i^2}{n_i}}} \sim t(n - g) \qquad \text{(T2.13)}$$

Thus the $1 - \alpha$ confidence interval of μ is exactly

$$a_0 + \sum_{i=1}^{g} a_i \bar{Y}_i \pm t_{n-g, 1-\alpha/2} \sqrt{\frac{1}{n-g} \sum_{i=1}^{g} (n_i - 1)s_i^2 \cdot \sum_{i=1}^{g} \frac{a_i^2}{n_i}}. \qquad \text{(T2.14)}$$

7.5 Estimation of SMCV and c^+-Probability

The estimation of SMCV relies on how samples are obtained in a study. When the groups are correlated, it is usually difficult to estimate the covariance among groups. In such a case, a good strategy is to obtain matched or paired samples (or subjects) and to conduct contrast analysis based on the matched samples. A simple example of matched contrast analysis is the analysis of paired difference of drug effects after and before taking a drug in the same patients. By contrast, another strategy is to not match or pair the samples and to conduct contrast analysis based on the unmatched or unpaired samples. A simple example of unmatched contrast analysis is the comparison of efficacy between a new drug taken by some patients and a standard drug taken by other patients. Methods of estimation for SMCV and c^+-probability in matched contrast analysis may differ from those used in unmatched contrast analysis.

7.5.1 Estimation of SMCV in Unmatched Samples

For unmatched contrast analysis, let us assume that t groups P_i's are independently and normally distributed. That is, P_i's are independent and $P_i \sim N(\mu_i, \sigma_i^2)$. The SMCV of contrast variable $V = \sum_{i=1}^{t} c_i P_i$, where $\sum_{i=1}^{t} c_i = 0$ is $\lambda = \sum_{i=1}^{t} c_i \mu_i / \sqrt{\sum_{i=1}^{t} c_i^2 \sigma_i^2}$. Consider an IID sample of size n_i, $Y_i = (Y_{i1}, Y_{i2}, \ldots, Y_{in_i})$ from the ith ($i = 1, 2, \ldots, t$) group P_i. Y_i's are independent. Let $\bar{Y}_i = \frac{1}{n_i} \sum_{j=1}^{n_i} Y_{ij}$, $s_i^2 = \frac{1}{n_i-1} \sum_{j=1}^{n_i} (Y_{ij} - \bar{Y}_i)^2$, $N = \sum_{i=1}^{t} n_i$, and MSE $= \frac{1}{N-t} \sum_{i=1}^{t} (n_i - 1)s_i^2$.

Applying Theorem 2 with $m = 1$, $a_0 = 0$, $a_i = c_i$, and $g = t$, we obtain the following results regarding the estimation of SMCV. When the t groups have unequal variance, the MLE of SMCV (from Formula T2.1) is:

$$\hat{\lambda}_{MLE} = \frac{\sum_{i=1}^{t} c_i \bar{Y}_i}{\sqrt{\sum_{i=1}^{t} \frac{n_i - 1}{n_i} c_i^2 s_i^2}} \tag{7.1}$$

The MM of SMCV (from T2.2) is:

$$\hat{\lambda}_{MM} = \frac{\sum_{i=1}^{t} c_i \bar{Y}_i}{\sqrt{\sum_{i=1}^{t} c_i^2 s_i^2}} \tag{7.2}$$

From T2.3, the asymptotical distribution of $\hat{\lambda}_{MLE}$ is $\hat{\lambda}_{MLE} \sim N(\lambda, \sigma_{\hat{\lambda}}^2)$ where

$$\sigma_{\hat{\lambda}}^2 = \frac{\sum_{i=1}^{t} (c_i^2 \sigma_i^2 / n_i)}{\sum_{i=1}^{t} c_i^2 \sigma_i^2} + \frac{1}{2} \frac{\left(\sum_{i=1}^{t} c_i \mu_i\right)^2 \sum_{i=1}^{t} (c_i^4 \sigma_i^4 / n_i)}{\left(\sum_{i=1}^{t} c_i^2 \sigma_i^2\right)^3}.$$

Thus when n_i's are all large, an estimate of $\sigma_{\hat{\lambda}}^2$ is

$$
\hat{\sigma}_{\hat{\lambda}}^2 = \frac{\sum_{i=1}^{t} \left(c_i^2 s_i^2 / n_i \right)}{\sum_{i=1}^{t} c_i^2 s_i^2} + \frac{1}{2} \frac{\left(\sum_{i=1}^{t} c_i \bar{Y}_i \right)^2 \sum_{i=1}^{t} \left(c_i^4 s_i^4 / n_i \right)}{\left(\sum_{i=1}^{t} c_i^2 s_i^2 \right)^3}
$$

and the $1 - \alpha$ confidence interval of SMCV λ is approximately

$$
\frac{\sum_{i=1}^{t} c_i \bar{Y}_i}{\sqrt{\sum_{i=1}^{t} \frac{n_i - 1}{n_i} c_i^2 s_i^2}} \pm z_{1-\alpha/2} \sqrt{\frac{\sum_{i=1}^{t} \frac{c_i^2 s_i^2}{n_i}}{\sum_{i=1}^{t} c_i^2 s_i^2} + \frac{1}{2} \frac{\sum_{i=1}^{t} \frac{c_i^4 s_i^4}{n_i}}{\left(\sum_{i=1}^{t} c_i^2 s_i^2 \right)^3} \cdot \left(\sum_{i=1}^{t} c_i \bar{Y}_i \right)^2}. \tag{7.3}
$$

In a situation with unequal variance but equal sample size r in each group, T2.4 becomes

$$
\frac{\sum_{i=1}^{t} c_i \bar{Y}_i}{\sqrt{\frac{1}{r} \sum_{i=1}^{t} c_i^2 s_i^2}} \sim t\left(\nu, \lambda \sqrt{r} \right) \text{ approximately, } \nu = (r-1) \frac{\left(\sum_{i=1}^{t} c_i^2 s_i^2 \right)^2}{\sum_{i=1}^{t} c_i^4 s_i^4}. \tag{7.4}
$$

Thus, whether the sample size is large or small, one can use the noncentral t-distribution in Formula 7.4 to get an approximate confidence interval as follows. Let $F_{t(\nu, b\lambda)}(\cdot)$ be the cumulative distribution function of noncentral $t(\nu, b\lambda)$ where $b = \sqrt{r}$ and ν is shown in Formula 7.4. Let T_{obs} be the observed value, namely $T_{\text{obs}} = \sum_{i=1}^{t} c_i \bar{Y}_i / \sqrt{\frac{1}{r} \sum_{i=1}^{t} c_i^2 s_i^2}$. Then we can find λ_L and λ_U such that $F_{t(\nu, b\lambda_L)}(T_{\text{obs}}) = 1 - \frac{\alpha}{2}$ and $F_{t(\nu, b\lambda_u)}(T_{\text{obs}}) = \frac{\alpha}{2}$; subsequently, (λ_L, λ_U) is approximately a $1 - \alpha$ confidence interval of SMCV λ. The MLE and MM estimates of SMCV are as shown in Formulas 7.1 and 7.2. An approximate unbiased estimate of SMCV (from Formula T2.5) is:

$$
\hat{\lambda}_{\text{AUE}} = \sqrt{\frac{2}{\nu}} \frac{\Gamma\left(\frac{\nu}{2} \right)}{\Gamma\left(\frac{\nu-1}{2} \right)} \frac{\sum_{i=1}^{t} c_i \bar{Y}_i}{\sqrt{\sum_{i=1}^{t} c_i^2 s_i^2}} \tag{7.5}
$$

When the t groups have equal variance, namely $\sigma_1^2 = \sigma_2^2 = \cdots = \sigma_t^2 = \sigma_e^2$, we commonly assume that the t groups P_i's are independently and normally distributed with $P_i \sim N(\mu_i, \sigma_e^2)$, which leads to the common model $Y_{ij} = \mu_i + \varepsilon_{ij}$ and $\varepsilon_{ij} \sim N(0, \sigma_e^2)$ in ANOVA. For a contrast variable $V = \sum_{i=1}^{t} c_i P_i$ where $\sum_{i=1}^{t} c_i = 0$, its mean is $\sum_{i=1}^{t} c_i \mu_i$ and its SMCV is $\lambda = \left(\sum_{i=1}^{t} c_i \mu_i \right) / \left(\sigma_e \sqrt{\sum_{i=1}^{t} c_i^2} \right)$.

As in classical textbooks, it is trivial to show that $T = \left(\sum_{i=1}^{t} c_i \bar{Y}_i - \sum_{i=1}^{t} c_i \mu_i \right) / \sqrt{\text{MSE} \cdot \sum_{i=1}^{t} c_i^2 / n_i} \sim t(N - t)$. The inference of contrast mean L is based on this central t-distribution. The $1 - \alpha$ confidence interval of L is:

$$T = \sum_{i=1}^{t} c_i \bar{Y}_i \pm t_{1-\alpha/2, N-t} \times \sqrt{\text{MSE} \cdot \sum_{i=1}^{t} c_i^2 / n_i}. \tag{7.6}$$

The UMVUE, MLE, and MM estimates of λ are, respectively (from T2.7–T2.9),

$$\hat{\lambda}_{\text{UMVUE}} = \frac{\sqrt{K}}{\sqrt{N - t}} \frac{\sum_{i=1}^{t} c_i \bar{Y}_i}{\sqrt{\text{MSE} \cdot \sum_{i=1}^{t} c_i^2}} \tag{7.7}$$

$$\hat{\lambda}_{\text{MLE}} = \sqrt{\frac{N}{N - t}} \frac{\sum_{i=1}^{t} c_i \bar{Y}_i}{\sqrt{\text{MSE} \cdot \sum_{i=1}^{t} c_i^2}} \tag{7.8}$$

$$\hat{\lambda}_{\text{MM}} = \frac{\sum_{i=1}^{t} c_i \bar{Y}_i}{\sqrt{\text{MSE} \cdot \sum_{i=1}^{t} c_i^2}} \tag{7.9}$$

where

$$K = 2 \cdot \left(\frac{\Gamma\left(\frac{n - t}{2}\right)}{\Gamma\left(\frac{n - t - 1}{2}\right)} \right)^2,$$

$N = \sum_{i=1}^{t} n_i$, and $\text{MSE} = \frac{1}{N-t} \sum_{i=1}^{t} (n_i - 1) s_i^2$. From T2.9, we have

$$T = \frac{\sum_{i=1}^{t} c_i \bar{Y}_i}{\sqrt{\text{MSE} \cdot \sum_{i=1}^{t} c_i^2 / n_i}} \sim \text{noncentral } t(\nu, b\lambda),$$

$$\text{where} \quad \nu = N - t \quad \text{and} \quad b = \sqrt{\frac{\sum_{i=1}^{t} c_i^2}{\sum_{i=1}^{t} c_i^2 / n_i}}. \tag{7.10}$$

Formula 7.10 can be used to construct the confidence interval. That is, let $F_{t(\nu, b\lambda)}(\cdot)$ be the cumulative distribution function of noncentral $t(\nu, b\lambda)$ and T_{obs} be the observed value of T. Then we can find λ_L and λ_U such that $F_{t(\nu, b\lambda_L)}(T_{\text{obs}}) = 1 - \frac{\alpha}{2}$ and $F_{t(\nu, b\lambda_u)}(T_{\text{obs}}) = \frac{\alpha}{2}$; subsequently, (λ_L, λ_U) is a $1 - \alpha$ confidence interval of SMCV λ.

Obviously, $\hat{\lambda}_{\text{UMVUE}}$, $\hat{\lambda}_{\text{MM}}$, and $\hat{\lambda}_{\text{MLE}}$ are all proportional to T. To simplify inferences of various SMCV estimates, the distribution of a random variable that is proportional to a noncentral t-distribution is called proportional noncentral t-distribution (or pnc t-distribution). That is, if $T \sim$ noncentral $t(p, \mu)$ and $W = aT$, then $W \sim$ pnc $t(p, \mu, a)$ [163;165;167]. Based on pnc t-distributions, $\hat{\lambda}_{\text{UMVUE}} \sim$ pnc $t(N - t, b\lambda, a_{\text{UMVUE}})$, $\hat{\lambda}_{\text{MM}} \sim$ pnc $t(N - t, b\lambda, a_{\text{MM}})$, and $\hat{\lambda}_{\text{MLE}} \sim$ pnc $t(N - t, b\lambda, a_{\text{MLE}})$, where $a_{\text{UMVUE}} = \sqrt{K/(N-t)} \cdot \sqrt{\sum_{i=1}^{t}(c_i^2/n_i)/\sum_{i=1}^{t} c_i^2}$, $a_{\text{MM}} = \sqrt{\sum_{i=1}^{t}(c_i^2/n_i)/\sum_{i=1}^{t} c_i^2}$, and $a_{\text{MLE}} = \sqrt{N/(N-t)} \cdot \sqrt{\sum_{i=1}^{t}(c_i^2/n_i)/\sum_{i=1}^{t} c_i^2}$. Hence the means of $\hat{\lambda}_{\text{UMVUE}}$, $\hat{\lambda}_{\text{MM}}$, and $\hat{\lambda}_{\text{MLE}}$ are λ, $(a_{\text{MM}}/a_{\text{UMVUE}})\lambda$, and $(a_{\text{MLE}}/a_{\text{UMVUE}})\lambda$ (i.e. λ, $\sqrt{\frac{N-t}{K}}\lambda$, and $\sqrt{\frac{N}{K}}\lambda$), respectively. The variances of $\hat{\lambda}_{\text{UMVUE}}$, $\hat{\lambda}_{\text{MM}}$, and $\hat{\lambda}_{\text{MLE}}$ are $a_{\text{UMVUE}}^2\sigma_T^2$, $a_{\text{MM}}^2\sigma_T^2$, and $a_{\text{MLE}}^2\sigma_T^2$, respectively, where

$$\sigma_T^2 = \frac{\nu}{\nu - 2} + \left(\frac{\nu}{\nu - 2} - \frac{\nu}{2}\left(\frac{\Gamma\left(\frac{\nu - 1}{2}\right)}{\Gamma\left(\frac{\nu}{2}\right)}\right)^2\right) b^2\lambda^2 \quad \text{and} \quad b = \sqrt{\frac{\sum_{i=1}^{t} c_i^2}{\sum_{i=1}^{t} c_i^2/n_i}}.$$

The previously described method of contrast variable and SMCV can be applied to any contrast in one-way ANOVA, although different contrasts may have different coefficients. After constructing a contrast variable, we can apply the same classifying rule in Table 7.2 to assess the strength of any contrast. For example, the effect in the ith group is commonly defined as $\tau_i = \mu_i - \mu_\bullet$ where $\mu_\bullet$ is the mean of μ_i's. The size of effect in the ith group can be assessed using a contrast variable

$$V_i = P_i - P_\bullet = P_i - \frac{1}{t}\sum_{k=1}^{t} P_k = \left(1 - \frac{1}{t}\right) P_i + \sum_{k \neq i}^{t}\left(-\frac{1}{t} P_k\right) = \sum_{k=1}^{t} c_k P_k$$

where $\quad c_k = \begin{cases} 1 - \frac{1}{t}, & \text{when } k = i \\ -\frac{1}{t}, & \text{when } k \neq i \end{cases}$.

Note that, corresponding to this contrast variable, $\mu_\bullet = \frac{1}{t}\sum_{k=1}^{t} \mu_k$ instead of $\frac{1}{N}\sum_{k=1}^{t} n_k\mu_k$ which implies $\bar{Y}_{\bullet\bullet} = \frac{1}{t}\sum_{k=1}^{t} \bar{Y}_{k\bullet}$ instead of $\frac{1}{N}\sum_{k=1}^{t} n_k\bar{Y}_{k\bullet}$.

7.5.2 Estimation of SMCV in Matched Samples

In matched contrast analysis, assume that we observe n independent samples $(Y_{1j}, Y_{2j}, \ldots, Y_{tj})$s from t groups Y_i's, where $i = 1, 2, \ldots, t$; $j = 1, 2, \ldots, n$. Then the jth observed value of a contrast $V = \sum_{i=1}^{t} c_i Y_i$ is $V_j = \sum_{i=1}^{t} c_i Y_{ij}$. Let $\bar{V}$ and s_V^2 be the sample mean and variance of contrast variable V. Suppose V_j's are IID with $V_j \sim N(\mu_V, \sigma_V^2)$.

If V_j here is treated as Y_{1j} in Theorem 2, then the matched contrast is a special case of the linear combination V in Theorem 2, with $g = 1$, $m = 1$, $a_0 = 0$, $a_1 = 1$, $n_1 = n$, $\bar{Y}_1 = \bar{V}$, and $s_1^2 = s_V^2$. Thus the UMVUE, MLE, and MM estimates of λ are, respectively (from T2.7–T2.9),

$$\hat{\lambda}_{\text{UMVUE}} = \frac{\sqrt{K}}{\sqrt{n-1}} \frac{\bar{V}}{\sqrt{s_V^2}} \quad \text{where} \quad K = 2 \cdot \left(\frac{\Gamma\left(\frac{n-1}{2}\right)}{\Gamma\left(\frac{n-2}{2}\right)} \right) \tag{7.11}$$

$$\hat{\lambda}_{\text{MLE}} = \sqrt{\frac{n}{n-1}} \frac{\bar{V}}{\sqrt{s_V^2}} \tag{7.12}$$

$$\hat{\lambda}_{\text{MM}} = \frac{\bar{V}}{\sqrt{s_V^2}} \tag{7.13}$$

From T2.9, we have

$$T = \frac{\bar{V}}{s_V / \sqrt{n}} \sim \text{noncentral } t\left(n-1, \sqrt{n}\lambda\right) \tag{7.14}$$

Based on this noncentral distribution of T, we can obtain the confidence interval of SMCV λ. That is, let $F_{t(\nu, b\lambda)}(\cdot)$ be the cumulative distribution function of noncentral $t(\nu, b\lambda)$ (where $b = \sqrt{n}$) and T_{obs} be the observed value of T. Then we can find λ_L and λ_U such that $F_{t(\nu, b\lambda_L)}(T_{\text{obs}}) = 1 - \frac{\alpha}{2}$ and $F_{t(\nu, b\lambda_u)}(T_{\text{obs}}) = \frac{\alpha}{2}$; subsequently, (λ_L, λ_U) is a $1 - \alpha$ confidence interval of SMCV λ. Note that traditional contrast analysis relies on point estimate $\bar{V}$, $1 - \alpha$ confidence interval $\bar{V} \pm T_{n-1, \frac{\alpha}{2}} \frac{s_V}{\sqrt{n}}$ of μ_V, and p-value based on $T = \sqrt{n} \frac{\bar{V}}{s_V} \sim$ central $t(n-1)$ under $H_0 : \mu_V = 0$, which can be applied to the inference of contrast mean. Let

$$a_{\text{UMVUE}} = \sqrt{\frac{2}{n(n-1)}} \frac{\Gamma\left(\frac{n-1}{2}\right)}{\Gamma\left(\frac{n-2}{2}\right)},$$

$a_{\text{MM}} = \frac{1}{\sqrt{n}}$, and $a_{\text{MLE}} = \frac{1}{\sqrt{n-1}}$. Then, $\hat{\lambda}_{\text{UMVUE}} \sim \text{pnc } t(n-1, \sqrt{n}\lambda, a_{\text{UMVUE}})$, $\hat{\lambda}_{\text{MLE}} \sim \text{pnc } t(n-1, \sqrt{n}\lambda, a_{\text{MLE}})$, and $\hat{\lambda}_{\text{MM}} \sim \text{pnc } t(n-1, \sqrt{n}\lambda, a_{\text{MM}})$. Therefore,

$$\text{E}(\hat{\lambda}_{\text{MLE}}) = \sqrt{\frac{n}{2}} \frac{\Gamma\left(\frac{n-2}{2}\right)}{\Gamma\left(\frac{n-1}{2}\right)} \lambda, \text{E}(\hat{\lambda}_{\text{MM}}) = \sqrt{\frac{n-1}{2}} \frac{\Gamma\left(\frac{n-2}{2}\right)}{\Gamma\left(\frac{n-1}{2}\right)} \lambda,$$

$$\text{Var}(\hat{\lambda}_{\text{UMVUE}}) = a_{\text{UMVUE}}^2 \sigma_T^2, \text{Var}(\hat{\lambda}_{\text{MLE}}) = a_{\text{MLE}}^2 \sigma_T^2, \text{ and } \text{Var}(\hat{\lambda}_{\text{MM}}) = a_{\text{MM}}^2 \sigma_T^2,$$

$$\text{where} \quad \sigma_T^2 = \frac{\nu}{\nu-2} + \left(\frac{\nu}{\nu-2} - \frac{\nu}{2} \left(\frac{\Gamma\left(\frac{\nu-1}{2}\right)}{\Gamma\left(\frac{\nu}{2}\right)} \right)^2 \right) b^2 \lambda^2 \quad \text{and} \quad b = \sqrt{n}.$$

7.5.3 Estimation of c^+-Probability

There are many approaches to the estimation of c^+-probability based on an observed sample. Here we describe three of them. The first is to use the relationship between SMCV and c^+-probability to transform an estimated SMCV value into the estimated value or limits of c^+-probability. The second is to use the strategy of exploring all possible combinations, as in estimating U-statistics to directly estimate c^+-probability. The third is to use a resampling method such as the bootstrap to directly get estimation of c^+-probability. For convenience, let us name transformed c^+-probability, combination c^+-probability, and bootstrap c^+-probability for the c^+-probability estimated by the first, second, and third approaches, respectively.

For the first approach, when a contrast variable has a normal distribution, the relationship between SMCV and c^+-probability is clear, as shown in Figure 7.2; thus we can transform any estimated SMCV value into a corresponding estimate of c^+-probability. We can even transform a confidence interval of SMCV into a confidence interval of c^+-probability. When the contrast variable does not have a normal distribution, it is difficult to directly obtain the one-to-one relationship between SMCV and c^+-probability. However, we can only obtain the upper or lower limits for c^+-probability in a situation in which the contrast variable has a unimodal distribution or a symmetric unimodal distribution with finite variance, as shown in Figure 7.2.

The second approach is similar to the direct method for calculating the Mann-Whitney-Wilcoxon U-statistic [157] when only two groups are involved in the contrast. When t ($t \geq 2$) groups are involved in a contrast, for every combination of t values from each of the t groups, respectively, we obtain a value for the contrast variable V; the proportion of the V values that are greater than zero in all the combinations is a combination c^+-probability. The third approach uses sampling with replacement. That is, for each run, sample one value with replacement from each of the t groups and obtain a corresponding value of contrast variable V based on the sampled values. The bootstrap c^+-probability is the proportion of the V values that are greater than zero in all runs. If the number of runs is set to be large enough, then the bootstrap c^+-probability should approximately equal the combination c^+-probability.

7.6 Contrasts in Multifactor ANOVA

7.6.1 A Contrast in General

In multifactor experiments with multiple levels in each factor, we use a random variable P_{kl} to represent the random values in the lth ($l = 1, \ldots, n_l$) level of the kth ($k = 1, \ldots, K$) factor. Suppose a set of coefficients c_l's represent a comparison for the levels in the kth factor. A contrast variable based on this set of coefficients is defined as $V = \sum_{l=1}^{n_l} c_l G_l$, where G_l is a random variable whose mean (and variance) equals the weighted mixture of means (and variances) of m combinations of factor levels containing the lth level of the kth factor.

Table 7.3. Random variables to represent values in each combination of levels in a two-factor experiment

Populations		Level 1	Level 2	$\cdots$	Level J	Pooled
					Factor 2	
	Level 1	P_{11}	P_{12}	$\cdots$	P_{1J}	$P_{1\bullet}$
	Level 2	P_{21}	P_{22}	$\cdots$	P_{2J}	$P_{2\bullet}$
Factor 1	$\vdots$	$\vdots$	$\vdots$	$\ddots$	$\vdots$	$\vdots$
	Level I	P_{I1}	P_{I2}	$\cdots$	P_{IJ}	$P_{I\bullet}$
	Pooled	$P_{\bullet1}$	$P_{\bullet2}$	$\cdots$	$P_{\bullet J}$	

Note: P_{ij} has mean μ_{ij} and variance σ_{ij}^2. $P_{i\bullet} = \sum_{j=1}^{J} \sqrt{w_{ij}}(P_{ij} - \mu_{ij}) + \sum_{j=1}^{J} w_{ij}\mu_{ij}$ where $\sum_{j=1}^{J} w_{ij} = 1$ for each i. $P_{\bullet j} = \sum_{i=1}^{I} \sqrt{w'_{ij}}(P_{ij} - \mu_{ij}) + \sum_{i=1}^{I} w'_{ij}\mu_{ij}$ where $\sum_{i=1}^{I} w'_{ij} = 1$ for each j.

Let us first look at a two-factor experiment in which the two factors have I and J levels, as shown in Table 7.3. Let us use a random variable P_{ij} to represent the random values in the combination of the ith level of the first factor and the jth level of the second factor. Assume P_{ij} has population mean μ_{ij} and variance σ_{ij}^2. A contrast variable for a comparison in the levels of factor 1 is

$$V = \sum_{i=1}^{I} c_{i\bullet} P_{i\bullet} \quad \text{where} \quad \sum_{i=1}^{I} c_{i\bullet} = 0 \tag{7.15}$$

$P_{i\bullet}$ is a random variable defined as $P_{i\bullet} = \sum_{j=1}^{J} \sqrt{w_{ij}}(P_{ij} - \mu_{ij}) + \sum_{j=1}^{J} w_{ij}\mu_{ij}$ where $\sum_{j=1}^{J} w_{ij} = 1$ for each i. Thus, its mean (and variance) equals the weighted mixture of means (and variances) of m combinations of levels in factors containing the ith level of factor 1, i.e., $\mu_{i\bullet} = \sum_{j=1}^{J} w_{ij}\mu_{ij}$, $\sigma_{i\bullet}^2 = \sum_{j=1}^{J} w_{ij}\sigma_{ij}^2$ and $\text{cov}(P_{i\bullet}, P_{i'\bullet}) = \sum_{j=1}^{J} (w_{ij}\text{cov}(P_{ij}, P_{i'j}))$.

The mean of V is

$$\mu_V = \sum_{i=1}^{I} c_{i\bullet}\text{E}(P_{i\bullet}) = \sum_{i=1}^{I} \left(c_{i\bullet} \sum_{j=1}^{J} w_{ij}\mu_{ij} \right) = \sum_{i=1}^{I}\sum_{j=1}^{J} (c_{i\bullet}w_{ij}\mu_{ij}).$$

The variance of V is

$$\sigma_V^2 = \text{Var}\left(\sum_{i=1}^{I} c_{i\bullet} P_{i\bullet} \right) = \sum_{i=1}^{I} c_{i\bullet}^2\text{Var}(P_{i\bullet}) + 2\sum_{i=1}^{I}\sum_{i'>i}^{I} \text{cov}(c_{i\bullet} P_{i\bullet}, c_{i'\bullet} P_{i'\bullet})$$

$$= \sum_{i=1}^{I} \left(c_{i\bullet}^2 \sum_{j=1}^{J} w_{ij}\sigma_{ij}^2 \right) + 2\sum_{i=1}^{I}\sum_{i'>i}^{I} \left(c_{i\bullet}c_{i'\bullet} \sum_{j=1}^{J} (w_{ij}\text{cov}(P_{ij}, P_{i'j})) \right)$$

Therefore, the standardized mean of V is

$$\lambda_V = \frac{\mu_V}{\sigma_V} = \frac{\sum_{i=1}^{I}\sum_{j=1}^{J}(c_{i\bullet}w_{ij}\mu_{ij})}{\sqrt{\sum_{i=1}^{I}\sum_{j=1}^{J}\left(c_{i\bullet}^2 w_{ij}\sigma_{ij}^2\right) + 2\sum_{i=1}^{I}\sum_{i'>i}^{I}\left(c_{i\bullet}c_{i'\bullet}\sum_{j=1}^{J}(w_{ij}\mathrm{cov}(P_{ij},P_{i'j}))\right)}}.$$

(7.16)

In a situation in which we are interested in the mixture of m levels ($m \leq J$) of factor 2 with equal weights (i.e., $w_{ij} = \frac{1}{m}$ if the jth level is of interest and $w_{ij} = 0$ otherwise), we have

$$\lambda_V = \frac{\sum_{i=1}^{I}\sum_{j=1}^{J}(c_{i\bullet}w_{ij}\mu_{ij})}{\sqrt{\sum_{i=1}^{I}\sum_{j=1}^{J}\left((c_{i\bullet}w_{ij})^2\sigma_{ij}^2\right) + 2\sum_{i=1}^{I}\sum_{i'>i}^{I}\left(\sum_{j=1}^{J}\mathrm{cov}(c_{i\bullet}w_{ij}P_{ij}, c_{i'\bullet}w_{i'j}P_{i'j})\right)}} \cdot \frac{1}{\sqrt{m}}.$$

(7.17)

As shown in Formula 7.15, the definition of a contrast variable for a comparison in levels of a factor is based on multiple random variables $P_{i\bullet}$'s which contain unknown parameter μ_{ij}'s. Thus, it is not easy to directly use Formula 7.15 to derive SMCV. With the observation in Formula 7.17, we can use the standardized mean of a linear combination of P_{ij}'s to calculate SMCV. Corresponding to the SMCV in Formula 7.17, we can construct a linear combination of P_{ij}'s as follows:

$$U = \sum_{i=1}^{I}\sum_{j=1}^{J} c_{i\bullet}w_{ij}P_{ij} = \sum_{i=1}^{I}\sum_{j=1}^{J} c_{ij}P_{ij} \quad \text{where} \quad c_{ij} = c_{i\bullet}w_{ij}$$

(7.18)

The standardized mean of the linear combination U is

$$\lambda_U = \frac{\mu_U}{\sigma_U} = \frac{\sum_{i=1}^{I}\sum_{j=1}^{J}(c_{i\bullet}w_{ij}\mu_{ij})}{\sqrt{\sum_{i=1}^{I}\sum_{j=1}^{J}\left((c_{i\bullet}w_{ij})^2\sigma_{ij}^2\right) + 2\sum_{i=1}^{I}\sum_{i'>i}^{I}\left(\sum_{j=1}^{J}\mathrm{cov}(c_{i\bullet}w_{ij}P_{ij}, c_{i'\bullet}w_{i'j}P_{i'j})\right)}}.$$

(7.19)

Because U directly consists of factor levels and is easier to handle than contrast variable V, we can work on U to get statistical inference on parameters from V based on Formulas 7.18 and 7.19. For convenience, we use the term contrast core to refer to the comparison (represented by $c_{i\bullet}$'s) embraced in a linear combination U and use the term core number to refer to the number of combinations with equal weights (i.e., m) [170].

In two-way ANOVA, we usually assume that P_{ij}'s are independent with equal variance σ_e^2. In such a case,

$$\lambda_V = \frac{\sum_{i=1}^{I} \sum_{j=1}^{J} (c_{i\bullet} w_{ij} \mu_{ij})}{\sqrt{\sum_{i=1}^{I} \sum_{j=1}^{J} (c_{i\bullet}^2 w_{ij}) \sigma_e^2}} = \frac{\sum_{i=1}^{I} \sum_{j=1}^{J} (c_{i\bullet} w_{ij} \mu_{ij})}{\sigma_e \sqrt{\sum_{i=1}^{I} c_{i\bullet}^2}} \quad \text{and} \quad \lambda_U = \frac{\sum_{i=1}^{I} \sum_{j=1}^{J} (c_{i\bullet} w_{ij} \mu_{ij})}{\sqrt{\sum_{i=1}^{I} \sum_{j=1}^{J} (c_{i\bullet} w_{ij})^2 \sigma_e^2}}.$$

In a situation in which we are interested in the mixture of m levels ($m \leq J$) of factor 2 with equal weights, we have

$$\lambda_V = \frac{\sum_{i=1}^{I} \sum_{j=1}^{J} (c_{i\bullet} w_{ij} \mu_{ij})}{\sqrt{\sum_{i=1}^{I} \sum_{j=1}^{J} (c_{i\bullet} w_{ij})^2 \sigma_e^2}} \cdot \frac{1}{\sqrt{m}} \tag{7.20}$$

Similarly, we can derive the contrast variable and its SMCV for the comparison of levels of factor 2. The above method of contrast variable and SMCV can be applied to any contrast in multifactor ANOVA, although different contrasts may have different coefficients. After constructing a linear combination of random variables that represent combinations of factor levels, we can apply the SMCV-based classifying rule in Table 7.2 [167] to assess the strength of any contrast. Again, in practice, the population values of SMCV are unknown, and we need to estimate them based on samples.

Consider a sample of size n_{ij}, and $Y_{ij} = (Y_{ij1}, Y_{ij2}, \ldots, Y_{ijn_{ij}})$ from the ijth ($i = 1, 2, \ldots, I; j = 1, 2, \ldots, J$) treatment combination P_{ij}. Y_{ij}'s are independent. Let

$$N = \sum_{i=1}^{I} \sum_{j=1}^{J} n_{ij}, \, n_{i\bullet} = \sum_{j=1}^{J} n_{ij}, \, \nu_e$$

$$= \sum_{i=1}^{I} \sum_{j=1}^{J} (n_{ij} - 1), \, K = 2 \cdot \left(\frac{\Gamma\left(\frac{\nu_e}{2}\right)}{\Gamma\left(\frac{\nu_e - 1}{2}\right)} \right)^2 \approx \nu_e - 1.48,$$

$$\bar{Y}_{ij\bullet} = \frac{1}{n_{ij}} \sum_{k=1}^{n_{ij}} Y_{ijk}, \, s_{ij}^2 = \frac{1}{n_{ij} - 1} \sum_{k=1}^{n_{ij}} (Y_{ijk} - \bar{Y}_{ij\bullet})^2,$$

and

$$\hat{\sigma}_e^2 = \text{MSE} = \frac{1}{\nu_e} \sum_{i=1}^{I} \sum_{j=1}^{J} (n_{ij} - 1) s_{ij}^2$$

$$= \frac{1}{\nu_e} \sum_{i=1}^{I} \sum_{j=1}^{J} \sum_{k=1}^{n_{ij}} \left(Y_{ijk} - \bar{Y}_{ij\bullet} \right)^2.$$

In this case, the estimate of traditional contrast L is $\hat{L} = \sum_{i=1}^{t} c_{ij} \bar{Y}_{ij\bullet}$.

In two-way ANOVA, a common model is $Y_{ijk} = \mu_{ij} + \varepsilon_{ijk}$, where $\varepsilon_{ijk} \sim N(0, \sigma_e^2)$ independently. Applying Theorem 2 in the situation of equal variance, we can readily obtain the following results. The UMVUE, MLE, and MM estimates of λ (from T2.6–2.8) are, respectively,

$$\hat{\lambda}_{\text{UMVUE}} = \frac{\sqrt{K}}{\sqrt{N - I \times J}} \frac{1}{\sqrt{m}} \frac{\sum_{i=1}^{I} \sum_{j=1}^{J} c_{ij} \bar{Y}_{ij\bullet}}{\sqrt{\text{MSE} \cdot \sum_{i=1}^{I} \sum_{j=1}^{J} c_{ij}^2}} \tag{7.21}$$

$$\hat{\lambda}_{\text{MLE}} = \frac{\sqrt{N}}{\sqrt{N - I \times J}} \frac{1}{\sqrt{m}} \frac{\sum_{i=1}^{I} \sum_{j=1}^{J} c_{ij} \bar{Y}_{ij\bullet}}{\sqrt{\text{MSE} \cdot \sum_{i=1}^{I} \sum_{j=1}^{J} c_{ij}^2}} \tag{7.22}$$

$$\hat{\lambda}_{\text{MM}} = \frac{1}{\sqrt{m}} \frac{\sum_{i=1}^{I} \sum_{j=1}^{J} c_{ij} \bar{Y}_{ij\bullet}}{\sqrt{\text{MSE} \cdot \sum_{i=1}^{I} \sum_{j=1}^{J} c_{ij}^2}} \tag{7.23}$$

From T2.9, we have

$$T = \frac{\sum_{i=1}^{I} \sum_{j=1}^{J} c_{ij} \bar{Y}_{ij\bullet}}{\sqrt{\text{MSE} \sum_{i=1}^{I} \sum_{j=1}^{J} \frac{c_{ij}^2}{n_{ij}}}} \sim \text{noncentral } t\,(\nu_e, b\lambda)\,, b = \sqrt{m \cdot \frac{\sum_{i=1}^{I} \sum_{j=1}^{J} c_{ij}^2}{\sum_{i=1}^{I} \sum_{j=1}^{J} c_{ij}^2 / n_{ij}}}. \tag{7.24}$$

Using this noncentral t-distribution, the confidence interval of SMCV λ can be constructed similarly as in one-way ANOVA. Considering $\hat{\lambda}_{\text{UMVUE}} = \frac{\sqrt{K}}{\sqrt{\nu_e}} bT$, $\hat{\lambda}_{\text{MM}} = bT$, $\hat{\lambda}_{\text{MLE}} = \frac{\sqrt{N}}{\sqrt{\nu_e}} bT$, and $T \sim$ noncentral $t(\nu_e, b\lambda)$, we have $\hat{\lambda}_{\text{UMVUE}} \sim \text{pnc } t\,(\nu_e, b\lambda, a_{\text{UMVUE}})$, $\hat{\lambda}_{\text{MM}} \sim \text{pnc } t\,(\nu_e, b\lambda, a_{\text{MM}})$, and $\hat{\lambda}_{\text{MLE}} \sim \text{pnc } t\,(\nu_e, b\lambda, a_{\text{MLE}})$, where $a_{\text{UMVUE}} = \frac{\sqrt{K}}{\sqrt{\nu_e}} b \approx \frac{\sqrt{\nu_e - 1.48}}{\sqrt{\nu_e}} b$, $a_{\text{MM}} = b$, and $a_{\text{MLE}} = \frac{\sqrt{N}}{\sqrt{\nu_e}} b$.

7.6.2 Effects of a Combination of Factor Levels

In ANOVA, we are usually interested in estimating $\tau_{ij} = \mu_{ij} - \mu_{\bullet\bullet}$. Here, the following contrast variable can be used to assess the strength of this effect:

$$V_{ij} = P_{ij} - P_{\bullet\bullet} = P_{ij} - \frac{1}{I \times J} \sum_{k=1}^{I} \sum_{l=1}^{J} P_{kl}$$

$$= \left(1 - \frac{1}{I \times J}\right) P_{ij} + \sum_{k \neq i}^{I} \sum_{l \neq j}^{J} \left(-\frac{1}{I \times J}\right) P_{kl} = \sum_{k=1}^{I} \sum_{l=1}^{J} c_{kl} P_{kl}$$

where

$$c_{kl} = \begin{cases} 1 - \frac{1}{I \times J}, & \text{when } k = i \text{ and } l = j \\ -\frac{1}{I \times J}, & \text{when } k \neq i \text{ or } l \neq j \end{cases}.$$

Clearly, $\sum_{k=1}^{I} \sum_{l=1}^{J} c_{kl} = 0$. Note that, corresponding to this contrast variable, $\mu_{\bullet\bullet} = \frac{1}{I \times J} \sum_{i=1}^{I} \sum_{j=1}^{J} \mu_{ij}$ instead of $\frac{1}{N} \sum_{i=1}^{I} \sum_{j=1}^{J} n_{ij} \mu_{ij}$, which implies that $\bar{Y}_{\bullet\bullet\bullet} = \frac{1}{I \times J} \sum_{i=1}^{I} \sum_{j=1}^{J} \bar{Y}_{ij\bullet}$ instead of $\frac{1}{N} \sum_{i=1}^{I} \sum_{j=1}^{J} \sum_{k=1}^{n_{ij}} Y_{ijk}$. The mean of V_{ij} is τ_{ij}, and the SMCV of V_{ij} is $\lambda_{ij} = \sum_{k=1}^{I} \sum_{l=1}^{J} c_{kl} \mu_{kl} \Big/ \sqrt{\sigma_e^2 \sum_{k=1}^{I} \sum_{l=1}^{J} c_{kl}^2}$. The estimate of τ_{ij} is $\hat{\tau}_{ij} = \sum_{k=1}^{I} \sum_{l=1}^{J} c_{kl} \bar{Y}_{kl\bullet} = \bar{Y}_{ij\bullet} - \bar{Y}_{\bullet\bullet\bullet}$ and the UMVUE estimate of τ_{ij} is

$$\hat{\lambda}_{ij} = \frac{\sqrt{K}}{\sqrt{v_e}} \frac{\sum_{k=1}^{I} \sum_{l=1}^{J} c_{kl} \bar{Y}_{kl\bullet}}{\sqrt{\text{MSE} \sum_{k=1}^{I} \sum_{l=1}^{J} c_{kl}^2}} = \frac{\sqrt{K}}{\sqrt{v_e}} b_{ij} T_{ij}, \quad \text{where } b_{ij} = \sqrt{\frac{\sum_{k=1}^{I} \sum_{l=1}^{J} c_{kl}^2}{\sum_{k=1}^{I} \sum_{l=1}^{J} c_{kl}^2 / n_{kl}}} \quad \text{and}$$

$$T_{ij} = \frac{\sum_{k=1}^{I} \sum_{l=1}^{J} c_{kl} \bar{Y}_{kl\bullet}}{\sqrt{\text{MSE} \sum_{k=1}^{I} \sum_{l=1}^{J} \frac{c_{kl}^2}{n_{kl}}}} \sim \text{noncentral } t\left(v_e, b_{ij} \lambda_{ij}\right).$$

We can use the distribution of T_{ij} to construct the confidence interval for λ_{ij}.

The relationship between $\hat{\tau}_{ij}$ and $\hat{\lambda}_{ij}$ is $\hat{\lambda}_{ij} = \sqrt{K/v_e} \Big/ \sqrt{\sum_{k=1}^{I} \sum_{l=1}^{J} c_{kl}^2} \cdot \frac{\hat{\tau}_{ij}}{\hat{\sigma}_e}$ for UMVUE estimate; it is $\hat{\lambda}_{ij} = 1 \Big/ \sqrt{\sum_{k=1}^{I} \sum_{l=1}^{J} c_{kl}^2} \cdot \frac{\hat{\tau}_{ij}}{\hat{\sigma}_e}$ for MM estimate. These relationships indicate that the value of commonly used standardized effect size $\frac{\hat{\tau}_{ij}}{\hat{\sigma}_e}$ does not have a consistent meaning in representing the strength of a comparison because of the existence of an adjusted coefficient $1 \Big/ \sqrt{\sum_{k=1}^{I} \sum_{l=1}^{J} c_{kl}^2}$ or $\sqrt{K/v_e} \Big/ \sqrt{\sum_{k=1}^{I} \sum_{l=1}^{J} c_{kl}^2}$, which varies in different contrasts.

7.6.3 Main Effects

In ANOVA, we are also interested in estimating $\tau_{i\bullet} = \mu_{i\bullet} - \mu_{\bullet\bullet} = \frac{1}{J} \sum_{j=1}^{J} \mu_{ij} - \frac{1}{I \times J} \sum_{i=1}^{I} \sum_{j=1}^{J} \mu_{ij}$ for the ith level of a factor. The main effect at the ith level of factor 1 is represented by the difference of the effect in the ith level and the combined effect in all the remaining levels of factor 1 across all the levels of factor 2. Thus this main effect can be assessed using the contrast variable $V_{i\bullet} = \sum_{k=1}^{I} c_{k\bullet} P_{k\bullet}$, where

$$c_{k\bullet} = \begin{cases} 1 - \frac{1}{I}, & \text{when } k = i \\ -\frac{1}{I}, & \text{when } k \neq i \end{cases}$$

and $P_{k\bullet}$ is a random variable whose mean (and variance) equals the weighted mixture of means (and variances) of all the J combinations of factor levels containing the kth level of factor 1, that is, $\mu_{k\bullet} = \sum_{j=1}^{J} \frac{1}{J} \mu_{kj}$ and $\sigma_{k\bullet}^2 = \sum_{j=1}^{J} \frac{1}{J} \sigma_{kj}^2$. This contrast variable is embraced in the following linear combination of factor levels with contrast core $c_{k\bullet}$'s and a core number of J.

$$
U_{i\bullet} = \frac{1}{J} \sum_{l=1}^{J} P_{il} - \frac{1}{I \times J} \sum_{k=1}^{I} \sum_{l=1}^{J} P_{kl}
$$

$$
= \sum_{l=1}^{J} \left(\frac{1}{J} - \frac{1}{I \times J} \right) P_{il} + \sum_{k \neq i}^{I} \sum_{l=1}^{J} \frac{-1}{I \times J} P_{kl} = \sum_{k=1}^{I} \sum_{l=1}^{J} c_{kl} P_{kl}
$$

where $c_{kl} = \begin{cases} \frac{1}{J} - \frac{1}{I \times J}, & \text{when } k = i \\ -\frac{1}{I \times J}, & \text{when } k \neq i \end{cases}$.

The SMCV of $V_{i\bullet}$ is

$$
\lambda_{i\bullet} = \frac{\sum_{k=1}^{I} c_{k\bullet} \mu_{k\bullet}}{\sqrt{\sigma_e^2 \sum_{i=1}^{I} c_{i\bullet}^2}} = \frac{\sum_{i=1}^{I} \sum_{j=1}^{J} c_{ij} \mu_{ij}}{\sqrt{\sigma_e^2 \sum_{i=1}^{I} \sum_{j=1}^{J} c_{ij}^2}} \cdot \frac{1}{\sqrt{J}} .
$$

Note that, corresponding to this contrast variable, $\mu_{i\bullet} = \frac{1}{J} \sum_{j=1}^{J} \mu_{ij}$ instead of $\frac{1}{n_{i\bullet}} \sum_{j=1}^{J} n_{ij} \mu_{ij}$; $\mu_{\bullet\bullet} = \frac{1}{I \times J} \sum_{i=1}^{I} \sum_{j=1}^{J} \mu_{ij}$ instead of $\frac{1}{N} \sum_{i=1}^{I} \sum_{j=1}^{J} n_{ij} \mu_{ij}$, which implies that $\bar{Y}_{i\bullet\bullet} = \frac{1}{J} \sum_{j=1}^{J} \bar{Y}_{ij\bullet}$ instead of $\frac{1}{n_{i\bullet}} \sum_{j=1}^{J} \sum_{k=1}^{n_{ij}} \bar{Y}_{ij\bullet}$; $\bar{Y}_{\bullet\bullet\bullet} = \frac{1}{I \times J} \sum_{i=1}^{I} \sum_{j=1}^{J} \bar{Y}_{ij\bullet}$ instead of $\frac{1}{N} \sum_{i=1}^{I} \sum_{j=1}^{J} \sum_{k=1}^{n_{ij}} Y_{ijk}$.

Thus we may estimate the mean $\tau_{i\bullet}$ and SMCV $\lambda_{i\bullet}$ of contrast variable $V_{i\bullet}$ as follows. The estimate of $\tau_{i\bullet}$ is $\hat{\tau}_{i\bullet} = \sum_{k=1}^{I} \sum_{l=1}^{J} c_{kl} \bar{Y}_{kl\bullet} = \bar{Y}_{i\bullet\bullet} - \bar{Y}_{\bullet\bullet\bullet}$, and the UMVUE estimate of $\lambda_{i\bullet}$ is

$$
\hat{\lambda}_{i\bullet} = \frac{\sqrt{K}}{\sqrt{\nu_e}} \frac{\sum_{k=1}^{I} \sum_{l=1}^{J} c_{kl} \bar{Y}_{kl\bullet}}{\sqrt{\text{MSE} \sum_{k=1}^{I} \sum_{l=1}^{J} c_{kl}^2}} \frac{1}{\sqrt{J}} = \frac{\sqrt{K}}{\sqrt{\nu_e}} b_{i\bullet} T_{i\bullet},
$$

where $b_{i\bullet} = \sqrt{\dfrac{J \sum_{k=1}^{I} \sum_{l=1}^{J} c_{kl}^2}{\sum_{k=1}^{I} \sum_{l=1}^{J} c_{kl}^2 / n_{kl}}}$

and $T_{i\bullet} = \dfrac{\sum_{k=1}^{I} \sum_{l=1}^{J} c_{kl} \bar{Y}_{kl\bullet}}{\sqrt{\text{MSE} \sum_{k=1}^{I} \sum_{l=1}^{J} \dfrac{c_{kl}^2}{n_{kl}}}} \sim$ noncentral $t\left(\nu_e, b_{i\bullet} \lambda_{i\bullet}\right)$.

We can use the distribution of $T_{i\bullet}$ to construct the confidence interval for $\lambda_{i\bullet}$. The relationship between $\hat{t}_{i\bullet}$ and $\hat{\lambda}_{i\bullet}$ is $\hat{\lambda}_{i\bullet} = 1/\sqrt{\sum_{k=1}^{I}\sum_{l=1}^{J} c_{kl}^2} \cdot \hat{t}_{i\bullet}/\hat{\sigma}_e \frac{1}{\sqrt{J}}$ for MM estimate of $\lambda_{i\bullet}$.

Similarly, the main effect at the jth level of factor 2 is represented by the difference of the effect in the jth level and the combined effect in all the remaining levels of factor 2 across all the levels of factor 1. This main effect can be investigated using contrast variable $V_{\bullet j}$: $V_{\bullet j} = \sum_{k=1}^{I} c_{\bullet l} P_{\bullet l}$, where

$$c_{\bullet l} = \begin{cases} 1 - \frac{1}{J}, & \text{when } l = j \\ -\frac{1}{J}, & \text{when } l \neq j \end{cases}$$

and $P_{\bullet l}$ is a random variable whose mean (and variance) equals the weighted mixture of means (and variances) of all the I combinations of factor levels containing the lth level of factor 2, that is, $\mu_{\bullet l} = \sum_{k=1}^{I} \frac{1}{I}\mu_{kl}$ and $\sigma_{\bullet l}^2 = \sum_{k=1}^{I} \frac{1}{I}\sigma_{kl}^2 = \sigma_e^2$. The corresponding linear combination of factor levels is

$$U_{\bullet j} = \frac{1}{I}\sum_{k=1}^{I} P_{kj} - \frac{1}{I \times J}\sum_{k=1}^{I}\sum_{l=1}^{J} P_{kl}$$

$$= \sum_{k=1}^{I}\left(\frac{1}{I} - \frac{1}{I \times J}\right) P_{kj} + \sum_{k=1}^{I}\sum_{l \neq j}^{J} \frac{-1}{I \times J} P_{kl} = \sum_{k=1}^{I}\sum_{l=1}^{J} c_{kl} P_{kl},$$

where

$$c_{kl} = \begin{cases} \frac{1}{I} - \frac{1}{I \times J}, & \text{when } l = j \\ -\frac{1}{I \times J}, & \text{when } l \neq j \end{cases}.$$

The mean of $V_{\bullet j}$ is $\tau_{\bullet j} = \frac{1}{I}\sum_{i=1}^{I}(\mu_{ij} - \mu_{\bullet\bullet}) = \mu_{\bullet j} - \mu_{\bullet\bullet}$, and the SMCV of $V_{\bullet j}$ is

$$\lambda_{\bullet j} = \frac{\sum_{l=1}^{J} c_{\bullet l}\mu_{\bullet l}}{\sqrt{\sigma_e^2 \sum_{l=1}^{J} c_{\bullet l}^2}} = \frac{\sum_{i=1}^{I}\sum_{j=1}^{J} c_{ij}\mu_{ij}}{\sqrt{\sigma_e^2 \sum_{i=1}^{I}\sum_{j=1}^{J} c_{ij}^2}} \cdot \frac{1}{\sqrt{I}}.$$

The mean $\tau_{\bullet j}$ for this main effect can be estimated by $\hat{t}_{i\bullet} = \sum_{k=1}^{I}\sum_{l=1}^{J} c_{kl}\bar{Y}_{kl\bullet} = \bar{Y}_{\bullet\bullet j} - \bar{Y}_{\bullet\bullet\bullet}$; the MM estimate of SMCV $\lambda_{\bullet j}$ for this main effect is $\hat{\lambda}_{\bullet j} = 1/\sqrt{\sum_{k=1}^{I}\sum_{l=1}^{J} c_{kl}^2} \cdot \hat{t}_{\bullet j}/\hat{\sigma}_e \frac{1}{\sqrt{I}}$ and the UMVUE estimate of SMCV $\lambda_{\bullet j}$ is

$$\hat{\lambda}_{\bullet j} = \frac{\sqrt{K}}{\sqrt{v_e}} \frac{\sum_{k=1}^{I}\sum_{l=1}^{J} c_{kl}\bar{Y}_{kl\bullet}}{\sqrt{\text{MSE} \sum_{k=1}^{I}\sum_{l=1}^{J} c_{kl}^2}} \frac{1}{\sqrt{I}}.$$

7.6.4 Interaction Effects

In ANOVA, the interaction effect between the ith level of the first factor and jth level of the second factor is traditionally defined as $\tau_{i \otimes j} = \mu_{ij} - (\mu_{\bullet\bullet} + \tau_{i\bullet} + \tau_{\bullet j}) = \mu_{ij} - \mu_{i\bullet} - \mu_{\bullet j} + \mu_{\bullet\bullet}$. This interaction effect $\tau_{i \otimes j}$ can be investigated using contrast variable $V_{i \otimes j}$:

$$V_{i \otimes j} = P_{ij} - P_{i\bullet} - P_{\bullet j} + P_{\bullet\bullet} = P_{ij} - \frac{1}{J} \sum_{l=1}^{J} P_{il} - \frac{1}{I} \sum_{k=1}^{I} P_{kj} + \frac{1}{I \times J} \sum_{k=1}^{I} \sum_{l=1}^{J} P_{kl}$$

$$= \left(1 - \frac{1}{I} - \frac{1}{J} + \frac{1}{I \times J}\right) P_{ij} + \sum_{l \neq j}^{J} \left(-\frac{1}{J} + \frac{1}{I \times J}\right) P_{il}$$

$$+ \sum_{k \neq i}^{I} \left(-\frac{1}{I} + \frac{1}{I \times J}\right) P_{kj} + \sum_{k \neq i}^{I} \sum_{l \neq j}^{J} \frac{1}{I \times J} P_{kl} = \sum_{k=1}^{I} \sum_{l=1}^{J} c_{kl} P_{kl}$$

$$\text{where } c_{kl} = \begin{cases} 1 - \dfrac{1}{I} - \dfrac{1}{J} + \dfrac{1}{I \times J}, & \text{when} \quad k = i, l = j \\[2mm] -\dfrac{1}{J} + \dfrac{1}{I \times J}, & \text{when} \quad k = i, l \neq j \\[2mm] -\dfrac{1}{I} + \dfrac{1}{I \times J}, & \text{when} \quad k \neq i, l = j \\[2mm] \dfrac{1}{I \times J}, & \text{when} \quad k \neq i, l \neq j \end{cases}.$$

The estimate of $\tau_{i \otimes j}$ is $\hat{\tau}_{i \otimes j} = \sum_{k=1}^{I} \sum_{l=1}^{J} c_{kl} \bar{Y}_{kl\bullet} = \bar{Y}_{ij\bullet} - \bar{Y}_{i\bullet\bullet} - \bar{Y}_{\bullet j\bullet} + \bar{Y}_{\bullet\bullet\bullet}$ and the UMVUE estimate of SMCV $\lambda_{i \otimes j}$ is

$$\hat{\lambda}_{i \otimes j} = \frac{\sqrt{K}}{\sqrt{v_e}} \frac{\sum_{k=1}^{I} \sum_{l=1}^{J} c_{kl} \bar{Y}_{kl\bullet}}{\sqrt{\text{MSE} \sum_{k=1}^{I} \sum_{l=1}^{J} c_{kl}^2}} = \frac{\sqrt{K}}{\sqrt{v_e}} b_{ij} T_{i \otimes j},$$

where

$$b_{i \otimes j} = \sqrt{\frac{\sum_{k=1}^{I} \sum_{l=1}^{J} c_{kl}^2}{\sum_{k=1}^{I} \sum_{l=1}^{J} c_{kl}^2 / n_{kl}}}.$$

We can use the distribution of $T_{i \otimes j}$ to construct the confidence interval for $\lambda_{i \otimes j}$. The relationship between $\hat{\tau}_{i \otimes j}$ and $\hat{\lambda}_{i \otimes j}$ is $\hat{\lambda}_{i \otimes j} = \sqrt{K/v_e} \Big/ \sqrt{\sum_{k=1}^{I} \sum_{l=1}^{J} c_{kl}^2} \cdot \hat{\tau}_{i \otimes j} / \hat{\sigma}_e$ for UMVUE estimate; it is $\hat{\lambda}_{i \otimes j} = 1 \Big/ \sqrt{\sum_{k=1}^{I} \sum_{l=1}^{J} c_{kl}^2} \cdot \hat{\tau}_{i \otimes j} / \hat{\sigma}_e$ for MM estimate of $\lambda_{i \otimes j}$.

All the above methods for two-way ANOVA can be readily extended to ANOVA with more than two factors. For example, in ANOVA that has three factors with I, J, and H levels, the main effect traditionally addressed by $\tau_{i\bullet\bullet} = \mu_{i\bullet\bullet} - \mu_{\bullet\bullet\bullet}$ in the ith level of the first factor can be addressed using contrast variable

$$
\begin{aligned}
V_{i\bullet\bullet} &= \frac{1}{J \times H} \sum_{l=1}^{J} \sum_{s=1}^{H} (P_{ils} - P_{\bullet\bullet\bullet}) \\
&= \frac{1}{J \times H} \sum_{l=1}^{J} \sum_{s=1}^{H} P_{ils} - \frac{1}{I \times J \times H} \sum_{k=1}^{I} \sum_{l=1}^{J} \sum_{s=1}^{H} P_{kls} \\
&= \sum_{l=1}^{J} \sum_{s=1}^{H} \left(\frac{1}{J \times H} - \frac{1}{I \times J \times H} \right) P_{ils} \\
&\quad + \sum_{k \neq i} \sum_{l=1}^{J} \sum_{s=1}^{H} \frac{-1}{I \times J \times H} P_{kl} = \sum_{k=1}^{I} \sum_{l=1}^{J} c_{kl} P_{kl}.
\end{aligned}
$$

Thus its coefficients are

$$
c_{kls} = \begin{cases} \dfrac{1}{J \times H} - \dfrac{1}{I \times J \times H}, & \text{when } k = i \\ -\dfrac{1}{I \times J \times H}, & \text{when } k \neq i. \end{cases}
$$

And its corresponding SMCV can be estimated using

$$
\hat{\lambda}_{i\bullet\bullet} = \frac{\sqrt{K}}{\sqrt{v_e}} \frac{\sum_{k=1}^{I} \sum_{l=1}^{J} \sum_{s=1}^{H} c_{kls} \bar{Y}_{kls\bullet}}{\sqrt{\text{MSE} \sum_{k=1}^{I} \sum_{l=1}^{J} \sum_{s=1}^{H} c_{kls}^2}} \frac{1}{\sqrt{J \times H}},
$$

where v_e is the degree of freedom in the three-way ANOVA and $K \approx v_e - 1.48$.

7.7 Case Studies and Simulation

7.7.1 A Simulation Study

Let us consider a simple comparison in which we want to compare the impact of siRNAs on cell viability of two cancer cell lines (i.e., a prostate cancer and a breast cancer cell line) with that of a stem cell line [167]. Let P_1, P_2, and P_3 denote the values of cell viability in the prostate cancer, breast cancer, and stem cell lines, respectively, and P_i ($i = 1, 2, 3$) has population mean μ_i and variance σ_i^2. In traditional contrast analysis, we can use a contrast $L_{\text{cancer}} = \frac{1}{2}\mu_1 + \frac{1}{2}\mu_2 - \mu_3$ to compare the effect of an siRNA on the two cancer cell lines with that of the stem cell line. In reality, the true values of μ_1, μ_2, μ_3 are unknown. The means in the samples from the three cell lines (i.e., $\bar{Y}_1$, $\bar{Y}_2$, $\bar{Y}_3$) are used to estimate the traditional contrast. That is, $\hat{L}_{cancer} = \frac{1}{2}\bar{Y}_1 + \frac{1}{2}\bar{Y}_2 - \bar{Y}_3$.

Consider two simulated experiments, one (i.e., experiment 1) with 100 observations (N = 100) in each cell line and the other (i.e., experiment 2) with 4 observations (N = 4) in each cell line. Suppose the measured values (in log2 scale) in the prostate cancer, breast cancer, and stem cell lines have normal distributions $N(13.06, 2.04^2)$, $N(12.94, 2.04^2)$, and $N(14, 2.04^2)$, respectively, for one siRNA (i.e., siRNA A1) in experiment 1 and another siRNA (i.e., siRNA A2) in experiment 2. The measured values in these three cell lines are from $N(13.48, 0.16^2)$, $N(13.35, 0.16^2)$, and $N(14, 0.16^2)$, respectively, for another two siRNAs in the two experiments (i.e., siRNA B1 in experiment 1 and siRNA B2 in experiment 2). The sample means in the three cell lines are 13.09, 12.82, and 14.04 for siRNA A1; 13.48, 13.37, and 14.01 for siRNA B1 (Figure 7.3A and B); 13.09, 12.71, and 13.95 for siRNA A2; and 13.53, 13.30, and 14.01 for siRNA B2 (Figure 7.3C and D), respectively.

The traditional contrast L_{cancer} is then estimated to be −1.09 (i.e., approximately an average 2-fold decrease in original scale) for siRNA A1 and −0.586 (i.e., approximately an average 1.5-fold decrease) for siRNA B1 in experiment 1, and −1.05 (i.e., approximately an average 2-fold decrease in original scale) for siRNA A2 and −0.59 (i.e., approximately an average 1.5-fold decrease) for siRNA B2 in experiment 2. In both experiments, the values of traditional contrast L_{cancer} suggest a larger average decrease of cell viability in the two cancer cell lines by siRNAs A1 and A2 than by siRNAs B1 and B2, which may lead to the conclusion that the two cancer cell lines are more different from the stem cell lines in siRNAs A1 and A2 than in siRNAs B1 and B2. However, Figure 7.3 (especially A and B) clearly reveals that, as a whole (or in a view of distributions), the two cancer cell lines are less different from the stem cell lines in siRNAs A1 and A2 than in siRNAs B1 and B2. Therefore, the conclusions reached using the values of traditional contrast are erroneous. These erroneous conclusions are produced not just by random chance; long as the sample size is large, the estimated value of the traditional contrast will approach to its population value, which leads to erroneous conclusions.

The p-values of traditional contrast L_{cancer} for siRNAs A1, A2, B1, and B2 are 0.00003, 0.55, 0, and 0.00006, respectively. Because both siRNA A1 and siRNA A2 have the same distributions for the three cell lines, the difference of effects between the two cancer cell lines and the stem cell lines in siRNA A2 should theoretically be the same as that in siRNA A1, as should be the relationship between siRNAs B2 and B1. However, the p-value for siRNA A1 is much less than that for siRNA A2, which may lead to the erroneous conclusion that the effects in the two cancer cell lines are much more different from those in the stem cell line in siRNA A1 than in siRNA A2. In addition, the p-value of the contrast for siRNA A1 is only half of the p-value for siRNA B2, which may again lead to the erroneous conclusion that the effects in

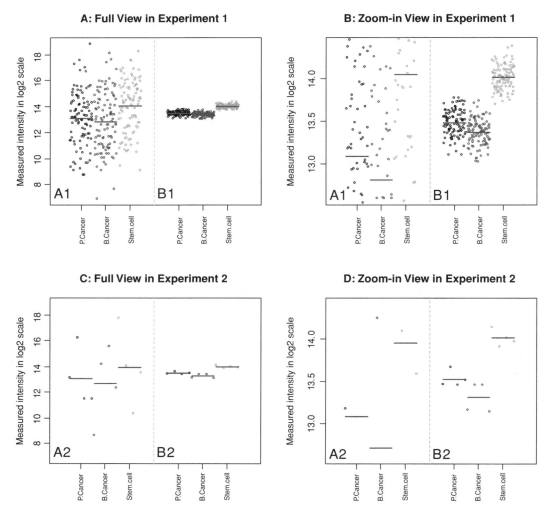

Figure 7.3   (See color insert following page 110.) The measured intensity in log2 scale of three cell lines, namely prostate cancer (P. Cancer), breast cancer (B. Cancer), and stem cell lines (Stem.cell), for siRNAs A1 and B1 in simulated experiment 1 and for siRNAs A2 and B2 in experiment 2. The measured values in the three cell lines are from normal distributions $N(13.06, 2.04^2)$, $N(12.94, 2.04^2)$, and $N(14, 2.04^2)$, respectively, for siRNAs A1 and A2, and are from $N(13.48, 0.16^2)$, $N(13.35, 0.16^2)$, and $N(14, 0.16^2)$, respectively, for siRNAs B1 and B2. The number of replicates in each cell line is 100 in experiment 1 and 4 in experiment 2. A red segment denotes a sample mean of measured value in a cell line for an siRNA. The two cancer cell lines have an average two-fold decrease in original scale compared with the stem cell line for siRNAs A1 and A2 and an average 1.5-fold decrease for siRNAs B1 and B2. *Source:* From Zhang [167].

the two cancer cell lines are much more different from those in the stem cell line in siRNA A1 than in siRNA B2. Theoretically, these erroneous conclusions can be produced as long as the difference of sample size in two experiments is large enough and thus they are produced not simply by chance.

For the simulated cancer study shown in Figure 7.3, we can use a contrast variable $V_{cancer} = \frac{1}{2}P_1 + \frac{1}{2}P_2 - P_3$ to compare the effects of an siRNA on the two cancer cell lines with those on the stem cell line. The SMCVs for siRNAs A1 and A2 were estimated to be −0.42 and −0.36, respectively, with both being close to the SMCV value of −0.40 for the distributions in which the two siRNAs are drawn. Based on the relationship between SMCV and c^+-probability, the SMCV values of −0.4, −0.42, and −0.36 are equivalent to the c^+-probability values of 0.345, 0.337, and 0.359, respectively. That is, if we randomly get one draw from each of the three cell lines, the chance that the average value of the two draws from the two cancer cell lines is less than the draw from the stem cell line is approximately 0.65 in both siRNAs A1 and A2. Based on the SMCV-based criteria in Table 7.2, both siRNAs A1 and A2 have a very weak effect.

Similarly, the SMCVs for siRNAs B1 and B2 were estimated to be −3.23 and −3.38, respectively; both are close to the SMCV value of −3 for the distributions in which the two siRNAs are drawn. Corresponding to the SMCV values of −3, −3.23, and −3.38, the c^+-probability values are 0.00135, 0.00062, and 0.00036, respectively. That is, if we randomly get one draw from each of the three cell lines, the chance that the average value of the two draws from the two cancer cell lines is less than the draw from the stem cell line is around 0.99865. Based on Table 7.2, this is a very strong effect for both siRNAs B1 and B2.

Therefore, the values of SMCV and c^+-probability of the four siRNAs correctly indicate that siRNAs A1 and A2 cause a very weak decrease and siRNAs B1 and B2 cause a very strong decrease in cell variability in the two cancer cell lines as compared with the stem cell line (Figure 7.3).

7.7.2 An Example of Matched Contrast Analysis

Fisher [50] used the data of Cushny and Peebles on the effect of optical isomers of hyoscyamine hydrobromide in producing sleep to display the misusage of unpaired t-test in paired samples. The same patients were used to test both isomers, laevo and dextro, for gaining additional hours of sleep. Thus the data were correlated and should be analyzed using the matched contrast approach. As demonstrated by Fisher [50], if we treated them as unpaired samples and ignored the correlation, we would have obtained a p-value of 0.0792 and subsequently would have declared that the mean difference is not significant, when in fact it is (Table 7.4). In addition to mean difference, a patient to be treated with one of the drugs tends to be more interested in these two questions (especially the second question): (i) what is the magnitude of difference in gained sleeping hours between the two isomers? and (ii) what is the probability that the number of gained sleeping hours for the patient treated with

Table 7.4. Examples of contrast analyses for the sleep study with matched samples

Contrast Coefficient	SMCV λ		Traditional Contrast L		
	Estimate	95% CI	Estimate	95% CI	p-Value for L = 0
As paired	1.174	(0.413, 2.127)	1.58	(0.70, 2.46)	0.0028
As unpaired	0.564	(−0.068, 1.233)	1.58	(−0.204, 3.363)	0.0792
As unpaired but capturing correlation	1.285	NA			

Note: CI, confidence interval; NA, not applicable.

laevo is greater than that of the patient treated with dextro? These two questions involve the use of information on distributions, not just means. Classical contrast analysis cannot appropriately answer these two questions.

Based on contrast variable $V_{sleep} = laevo - dextro$, we can use SMCV to address the question about the strength of contrast (i.e., magnitude of difference between the two isomers). The SMCV of the paired difference is estimated to be 1.174 using UMVUE and 1.285 using MM; thus the magnitude of the paired difference is large based on the criteria listed in Table 7.2. For SMCV, if we treated paired samples as unpaired and ignored correlations, we similarly would obtain misleading results. Based on the general definition of SMCV (i.e., $\lambda = (\mu_1 - \mu_2)/\sqrt{\sigma_1^2 + \sigma_2^2 - 2\sigma_{12}}$ in this case), when we ignore correlation, we underestimate SMCV if the correlation is positive and overestimate SMCV if the correlation is negative. In the sleeping drug case, the correlation is positive; therefore, ignoring the correlation leads to underestimation of the magnitude of difference between the two isomers (i.e., 0.564 instead of 1.174 or 1.285). On the other hand, if we treat the observations as unpaired but capture the covariance using the MM estimate, we would obtain the estimated SMCV to be 1.285, which is the same as the value of the MM estimate in a situation in which the observations are treated as paired samples. However, it may be nontrivial to obtain the confidence interval for this MM estimate of SMCV if we treat the observations as correlated unpaired samples.

Based on a contrast variable, we can also use c^+-probability to address the question regarding the probability that a patient will gain more sleeping hours by taking laevo instead of dextro isomers. After we transform SMCV estimates into c^+-probability based on the relationship between SMCV and c^+-probability in Theorem 1, the c^+-probability for the paired difference is estimated to be 0.88 for a normal distribution, to be no less than 0.84 for any symmetric unimodal distribution, and to be no less than 0.68 for any unimodal distribution. The observed c^+-probability (i.e., the proportion of observed values of the paired difference being greater than zero) is 0.9, which is close to 0.88.

Table 7.5. Empathy scores for boys and girls in grades 6, 9, and 12

Empathy	Score	Grade 6	Grade 9	Grade 12
Girls	Observed	35, 30, 39, 42, 48,	54, 58, 46, 65, 48,	65, 67, 62, 54, 58,
	values	37, 39, 46, 42, 41	55, 59, 64, 56	55, 58, 53, 50, 60
	Means	39.9	56.1	58.2
Boys	Observed	34, 28, 30, 35, 25,	55, 45, 57, 53, 51,	62, 67, 55, 59, 58,
	values	39, 35, 38, 27, 26	63, 47, 60, 50, 46	68, 53, 56, 54
	Means	31.7	52.7	59.1

MSE: 31.586; degrees of freedom ν_e: 52.

7.7.3 An Illustrative Example for Contrast Variables in ANOVA

Here I construct an illustrative example to demonstrate how contrast variables work in ANOVA. In this example, empathy scores for girls and boys in grades 6, 9, and 12 are obtained, with the data listed in Table 7.5 and displayed in Figure 7.4.

This is an example of ANOVA with two factors, gender and grade. The factor gender has two levels, girls and boys, and the factor grade has three levels, grades 6, 9, and 12. That is, $I = 2$ and $J = 3$. Let random variables P_{11}, P_{12}, P_{13}, P_{21}, P_{22}, and P_{23} represent the empathy scores of the girls in grades 6, 9, and 12 and the boys in grades 6, 9, and 12, respectively. The main effect for the girls can be assessed using a contrast variable $V_{1\bullet} = \frac{1}{2}P_{1\bullet} - \frac{1}{2}P_{2\bullet}$, where $P_{1\bullet}$ is a random variable for the girls whose mean (and variance) equals the equally weighted mixture of means (and variances) of three combinations of levels (i.e., girls in grades 6, 9, and 12) and $P_{2\bullet}$ is a similar random variable for the boys. The mean and SMCV of $V_{1\bullet}$ can be obtained

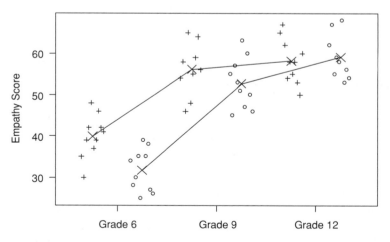

Figure 7.4   Displaying data in an illustrative example about empathy scores of girls and boys in grades 6, 9, and 12. The plus (+) symbols and circles represent the scores for girls and boys, respectively. An × represent the mean of scores in a treatment level.

using the linear combination $U_{1\bullet} = \frac{1}{6}P_{11} + \frac{1}{6}P_{12} + \frac{1}{6}P_{13} - \frac{1}{6}P_{21} - \frac{1}{6}P_{22} - \frac{1}{6}P_{23}$. For $U_{1\bullet}$, the coefficient set is $\left(\frac{1}{6}, \frac{1}{6}, \frac{1}{6}, -\frac{1}{6}, -\frac{1}{6}, -\frac{1}{6}\right)$, the contrast core is $\left(\frac{1}{2}, -\frac{1}{2}\right)$, and the core number is 3. Thus the SMCV of V is estimated as follows:

$$
\hat{\lambda}_{1\bullet} = \frac{\sqrt{\nu_e - 1.5}}{\sqrt{\nu_e}} \frac{\sum\limits_{k=1}^{I}\sum\limits_{l=1}^{J} c_{ij}\bar{Y}_{ij\bullet}}{\sqrt{\text{MSE}\sum\limits_{k=1}^{I}\sum\limits_{j=1}^{J} c_{ij}^2}} \frac{1}{\sqrt{J}}
$$

$$
= \frac{\sqrt{52 - 1.5}}{\sqrt{52}} \frac{\frac{1}{6} \times 39.9 + \frac{1}{6} \times 56.1 + \frac{1}{6} \times 58.2 - \frac{1}{6} \times 31.7 - \frac{1}{6} \times 52.7 - \frac{1}{6} \times 59.1}{\sqrt{31.586 \times \left[\frac{1^2}{6} + \frac{1^2}{6} + \frac{1^2}{6} + \left(-\frac{1}{6}\right)^2 + \left(-\frac{1}{6}\right)^2 + \left(-\frac{1}{6}\right)^2\right]}}
$$

$$
\times \frac{1}{\sqrt{3}} = 0.442,
$$

and the confidence interval of $\lambda_{1\bullet}$ can be obtained using

$$
T = \frac{\sum\limits_{i=1}^{I}\sum\limits_{j=1}^{J} c_{ij}\bar{Y}_{ij\bullet}}{\sqrt{\text{MSE}\sum\limits_{i=1}^{I}\sum\limits_{j=1}^{J} \frac{c_{ij}^2}{n_{ij}}}} \sim \text{noncentral } t\left(\nu_e, b\lambda_{1\bullet}\right),
$$

$$
b = \sqrt{J \cdot \frac{\sum\limits_{i=1}^{I}\sum\limits_{j=1}^{J} c_{ij}^2}{\sum\limits_{i=1}^{I}\sum\limits_{j=1}^{J} c_{ij}^2 / n_{ij}}} = \sqrt{3 \cdot \frac{6 \times \left(\frac{1}{6}\right)^2}{\left(\frac{1}{6}\right)^2 \left(\frac{1}{10} \times 4 + \frac{1}{9} \times 2\right)}} = 5.3785.
$$

Similarly, we applied the method of contrast variable to investigate other main effects at levels of either factor and to explore linear and quadratic relationships. The results are listed in Table 7.6.

The results in Table 7.6 show that the SMCV of the contrast variable for girl main effect is 0.442 and thus the main effect of girls is very weak; so is the main effect of boys. The average magnitude of gender effects is 0.442, which is a very weak effect. In other words, the gender factor has a very weak effect, or the overall difference between girls and boys is very weak. For each grade level, the difference between girls and boys is fairly moderate (i.e., SMCV = 1.017) at grade 6, very weak (i.e, SMCV = 0.423) at grade 9, and extremely weak (i.e., SMCV = −0.113) at grade 12. The SMCVs of contrast variables for main effects at grades 6, 9, and 12 are −2.968, 1.028, and 1.940, respectively, which indicates that the strengths of main effects at grades 6, 9, and 12 are strong, fairly moderate, and fairly strong, respectively. The average magnitude of grade effects is 1.979; thus the grade factor has a fairly strong effect.

Table 7.6. Contrast analysis using SMCV and traditional contrast L for an illustrative example about empathy scores of girls and boys in grades 6, 9, and 12

Contrasts (Coefficients)	SMCV	Strength Types	95% CI of SMCV	Estimated Value of L	p-Value of Testing $L = 0$
Girl $\tau_{1\bullet}$: $\left(\frac{1}{6}, \frac{1}{6}, \frac{1}{6}, -\frac{1}{6}, -\frac{1}{6}, -\frac{1}{6}\right)$	0.442	Very weak	(0.072, 0.821)	1.78	0.019
Boy $\tau_{2\bullet}$: $\left(-\frac{1}{6}, -\frac{1}{6}, -\frac{1}{6}, \frac{1}{6}, \frac{1}{6}, \frac{1}{6}\right)$	−0.442	Very weak	(−0.821, −0.072)	−1.78	0.019
Average magnitude of gender effects	0.442	Very weak			
Grade 6 $\tau_{\bullet 1}$: $\left(\frac{1}{3}, -\frac{1}{6}, -\frac{1}{6}, \frac{1}{3}, -\frac{1}{6}, -\frac{1}{6}\right)$	−2.968	Strong	(−3.733, −2.279)	−13.8	0
Grade 9 $\tau_{\bullet 2}$: $\left(-\frac{1}{6}, \frac{1}{3}, -\frac{1}{6}, -\frac{1}{6}, \frac{1}{3}, -\frac{1}{6}\right)$	1.028	Fairly moderate	(0.548, 1.529)	4.8	3×10^{-5}
Grade 12 $\tau_{\bullet 3}$: $\left(-\frac{1}{6}, -\frac{1}{6}, \frac{1}{3}, -\frac{1}{6}, -\frac{1}{6}, \frac{1}{3}\right)$	1.940	Fairly strong	(1.377, 2.549)	9.0	1×10^{-11}
Average magnitude of grade effects	1.979	Fairly strong			
Linear: $(-1, 0, 1, -1, 0, 1)$	2.834	Strong	(2.162, 3.578)	45.7	0
Quadratic: $(-1, 2, -1, -1, 2, -1)$	1.028	Fairly moderate	(0.548, 1.529)	28.7	3×10^{-5}

The SMCV is estimated to 2.834 for the contrast variable representing the linear relationship between grade levels and empathy scores, which indicates that this linear relationship is strong. Judged from the coefficients, the contrast variable representing the linear relationship also denotes the difference of empathy scores between grades 12 and 6. The SMCV is estimated to 1.028 for the contrast variable representing the quadratic relationship between grade levels and empathy scores, which indicates that this quadratic relationship is fairly moderate. Judged from the coefficients, the contrast variable representing the quadratic relationship also denotes the main effect at grade 9, namely the difference between grade 9 and the average of grades 6 and 12. Judged by observing the data displayed in Figure 7.4, all the preceding conclusions about the strength of main effects and other contrasts obtained using SMCV are reasonable.

If we use the values of contrast means (namely traditional contrasts) and associated p-values, we conclude that all the contrast means listed in Table 7.6 are significant except the differences between girls and boys at grades 9 and 12, respectively, and we can hardly obtain useful information about the strength of comparison. One more case is that we know the contrast variables with coefficients $(-1, 2, -1, -1, 2, -1)$ and $\left(-\frac{1}{6}, \frac{1}{3}, -\frac{1}{6}, -\frac{1}{6}, \frac{1}{3}, -\frac{1}{6}\right)$, respectively, can both represent the quadratic relationship between empathy scores and grade levels. However, the estimated values of traditional contrast for these two contrasts are very different (i.e., 28.7 and 4.8, respectively). By contrast, the values of SMCV are the same (i.e., 1.028) for both contrasts (Table 7.6), which also indicates that the results reached using SMCV are reasonable, better than those reached using a traditional contrast.

7.7.4 An Illustrative Example for Phenotypic Effects of an siRNA in Multifactor ANOVA

Figure 7.5 shows the measured phenotypic effects of an siRNA in experiments with two factors: cell line (three levels, cell lines 1–3) and treatment (two levels, with a drug and without a drug). In the illustrative experiments shown in Figure 7.5A–E, cell lines have no impact on the measured phenotypic effects. In other words, the phenotypic effects in any cell line treated with a drug come from a normal distribution $N(2, 0.245)$; those in any cell line treated without a drug come from another normal distribution $N(3.35, 0.245)$. For simplicity, the data in Figure 7.5A through E are generated so that the sample mean equals 2 in any cell line treated with the drug and 3.35 without the drug. The sample standard deviation equals 0.245 in each combination level of the two factors. Consequently, the drug effects are the same in each cell line.

The core interest is the drug effect, that is, the comparison of the phenotypic effects in cells treated with the drug and those without the drug. Thus the contrast core is the difference between with the drug and without the drug (i.e., drug − noDrug). For the experiments in Figure 7.5A and D, the contrast variable to address the core interest across the first two cell lines can be investigated using the linear combination of factor levels: $U_1 = (\text{drug.cell1} - \text{noDrug.cell1} + \text{drug.cell2} - \text{noDrug.cell2})/2$, where drug.Cell1 and noDrug.Cell1 denote the phenotypic effects in cell line 1 treated with the drug and without the drug, respectively. Similar notations are applied to other cell lines. The coefficients are $c_{1j} = \frac{1}{2}$ and $c_{2j} = -\frac{1}{2}$ for $j = 1, 2$. The core number is 2 in this contrast (i.e., $m = 2$). Similarly, for the experiments in Figure 7.5B, E, and F, the contrast variable to address the core interest across the three cell lines can be investigated using: $U_2 = (\text{drug.cell1} - \text{noDrug.cell1} + \text{drug.cell2} - \text{noDrug.cell2} + \text{drug.cell3} - \text{noDrug.cell3})/3$. Correspondingly, $c_{1j} = \frac{1}{3}$ and $c_{2j} = -\frac{1}{3}$ for $j = 1, 2, 3$, and the core number is 3 (i.e., $m = 3$).

Because the data about phenotypic effects in one cell line are just a repeat of those in another cell line, the drug effects are the same in each cell line. Thus the size of drug effects should be the same regardless of how many cell lines are involved in the study. The size of drug effects is represented by the magnitude of difference between phenotypic effects in cells treated with the drug and those without the drug. The contrast analysis to address for this difference (i.e., drug effect) should obtain the same or at least similar values when one, two, or three cell lines are involved. However, based on traditional contrast analysis, the p-values for the contrasts addressing drug effects are 0.112245, 0.018191, and 0.003245 in one, two, and three cell lines, respectively (Figure 7.5C, A, and B, respectively). That is, the first is nonsignificant at the level of 0.05, the second is significant at the level of

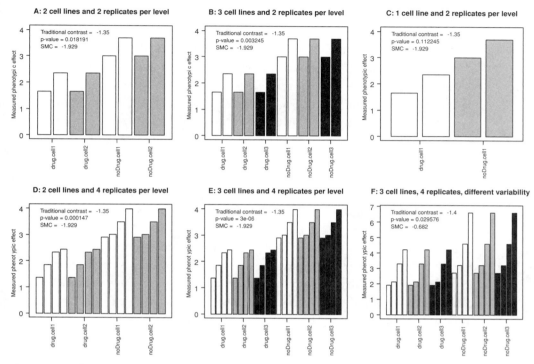

Figure 7.5   Phenotypic effects of an siRNA in illustrative experiments with three cell lines (cell lines 1–3) and two treatments (with a drug and without a drug). The core interest is the difference between phenotypic effects caused by the siRNA in cells treated with a drug and without a drug. The measured values of phenotypic effects of the siRNA with (or without) the drug in one cell line are the same as those in another cell line. Traditional contrast, p-value, and SMCV denote the sample mean difference of phenotypic effects between with the drug and without the drug, the p-value for testing the difference being zero, and standardized mean of contrast variable estimated using the MM method, respectively. *Source:* From Zhang [170].

0.05 and nonsignificant at the level of 0.01, and the third is significant at both levels, which leads to very different conclusions about the same size of drug effects. By contrast, using contrast variables, the SMCV is the same (i.e., −1.929) regardless of whether one, two, or three cell lines are involved (Figure 7.5C, A, and B, respectively), which correctly indicates that the size of drug effects is the same regardless of the number of cell lines used in the experiments.

The size of drug effects is the same when the sample size increases (i.e., when the number of replicates per combination of factor levels increases from two to four) in Figure 7.5A through E. However, the p-value in traditional contrast analysis decreases dramatically from 0.018191 to 0.000147 if two cell lines are used (Figure 7.5A and D) and from 0.003245 to 0.000003 if three cell lines are used (Figure 7.5B and E) when the sample size increases, which again generates different conclusions about the drug effects. The SMCV is still the same when the sample size increases, which correctly indicates that the size of drug effects is the same regardless of the sample size used in the studies.

One may note that the value of traditional contrast is the same (i.e., −1.35) in Figure 7.5A through E. However, one issue with the traditional contrast is that it is sensitive to the multiplication of a constant to its coefficients. For example, if each coefficient of the contrast in Figure 7.5B is doubled, then the value of the corresponding traditional contrast is also doubled to be −2.7. By contrast, the SMCV will stay the same as −1.929 because the constant 2 in the numerator of SMCV will be canceled out by the same constant in the denominator of SMCV. One more serious issue is that it is well known that the value of traditional contrast cannot capture data variability and may lead to misleading results. For example, the values of traditional contrast indicate that the magnitude of difference in the situation shown in Figure 7.5E is weaker than that of the situation shown in Figure 7.5F. However, the data clearly reveal that the magnitude of difference in the situation shown in Figure 7.5E is stronger than that of the situation shown in Figure 7.5F because of the consistently lower values in cells treated with the drug than without the drug in Figure 7.5E. By contrast, the values of SMCV correctly indicate that the drug effect is fairly strong in Figure 7.5E and weak in Figure 7.5F.

7.8 Discussion and Conclusions

Group comparison is a common practice in statistical analysis, especially for hit selection and quality control in genome-scale RNAi screens. The concepts of contrast variable and associated terms of SMCV and c^+-probability can link together the most commonly used probabilistic index for effect sizes, such as $\Pr(X > Y)$, and the most commonly used ratios of mean difference to variability, such as Cohen's d. A contrast variable can provide both a probabilistic meaning and an index of signal-to-noise ratio to interpret the strength of a comparison, which offers us a strong base to classify the strength of a comparison, as shown in Table 7.2. SMCV and c^+-probability also give interpretations to both Cohen's and McLean's criteria [33;104]. The contrast variable, SMCV, and c^+-probability work effectively and consistently for either relationship or group comparison in either independent or correlated situations and in either two or more than two groups. Treatment effect, main effect, interaction effect, linear relationship, quadratic relationship, and any other contrasts can all be addressed consistently using contrast variables. The examples in this chapter show that the results reached using contrast variables and the classifying rule are sensible and matched with observations and intuitions from the data. Therefore, the contrast variable, SMCV, c^+-probability, and SMCV-based classifying rule may have a broad utility for group comparison.

As a caveat, the SMCV-based classifying rule relies on the population values of SMCV. Because the population value of SMCV is usually unknown, sampling variability will play a role in the use of SMCV-based criteria, especially when the sample size is small. Both UMVUE and MLE estimates work best when the data are normally distributed, although MLE estimate and related confidence interval may work reasonably well with a large sample size, even when the assumption of normality is violated. The MM estimate of SMCV does not assume normal distributions; however, the distribution of the MM estimate can be complicated. Given all the previously mentioned features of SMCV estimates, along with the fact that the SMCV-based classifying rule works best in situations in which the data are normally distributed, it is a good idea to check normality and/or to transform the data to be nearly normal before applying the SMCV methods.

The concepts and theorems for group comparisons discussed in this chapter have been developed from a statistical theoretical basis. They provide a foundation for deriving statistical methods for data analysis in RNAi screens. In the next chapter, I elaborate how to derive these statistical methods for hit selection and quality control in genome-scale RNAi screens.

Statistical Methods for Assessing the Size of siRNA Effects

The size of an siRNA effect is represented by the magnitude of difference between the siRNA and a negative reference. Traditionally, mean difference (or, equivalently, average fold change in log scale), along with p-value of testing mean difference, has been used to indicate siRNA effects. However, as a statistical parameter, mean difference does not contain any information about data variability, cannot effectively measure the magnitude of difference between two groups, and thus cannot be used to assess siRNA effects successfully. Recently, SSMD and d^+-probability have been proposed for the comparison of two groups [161;162;165] and have been extended to multigroup comparisons [163;167;170]. SSMD is a special case of SMCV when only two groups are involved in a comparison. Thus, given the concepts and theorems regarding contrast variable, SMCV, and c^+-probability presented in Chapter 7, I explore the use of SSMD and d^+-probability for assessing the size of siRNA effects in this chapter. Specifically, I first present the concepts of SSMD and d^+-probability along with their relationship in Section 8.1 and the estimation of SSMD in Section 8.2. Standardized mean difference has been used for measuring the magnitude of difference, and classical t-statistic has been used for selecting hits in screens with replicates. Both look similar to SSMD. Therefore, I compare SSMD with standardized mean difference and classical t-statistic in Section 8.3. Given the concepts of SSMD and d^+-probability, as well as their estimation, I explore how to use SSMD to rank siRNAs in Section 8.4, how to control FPRs and FNRs in Section 8.5, and how to control FDRs and FNDRs in SSMD-based decision rules for selecting hits in genome-scale RNAi screens in Section 8.6. Finally, I derive analytic methods addressing off-target effects for hit selection in experiments with multiple siRNAs against a gene in Section 8.7 and present conclusions in Section 8.8.

8.1 SSMD and d^+-Probability

Suppose we are interested in the comparison of two groups. The first group has a distribution F_1 with mean μ_1 and variance σ_1^2 and the second group has a distribution

F_2 with mean μ_2 and variance σ_2^2. The covariance between these two groups is σ_{12}. Let random variables P_1 and P_2 denote random values in these two groups, respectively. SSMD (denoted as β) is defined as the ratio of mean to standard deviation of the difference between two groups, namely

$$\beta = \frac{\mu_D}{\sigma_D} = \frac{\mu_1 - \mu_2}{\sqrt{\sigma_1^2 + \sigma_2^2 - 2\sigma_{12}}}. \tag{8.1}$$

If the two groups are independent,

$$\beta = \frac{\mu_1 - \mu_2}{\sqrt{\sigma_1^2 + \sigma_2^2}}. \tag{8.2}$$

If the two independent groups have equal variance σ^2, then

$$\beta = \frac{\mu_1 - \mu_2}{\sqrt{2\sigma^2}}. \tag{8.3}$$

The probability that the random variable D representing the difference between two groups obtains positive values is d^+-probability. That is, d^+-probability $=$ $\Pr(D > 0)$. If we get a random draw from each group and calculate the sampled value of the difference between the two random draws, d^+-probability is the chance that the sampled values of the difference are greater than zero when the random draw process is repeated infinite times.

Like SSMD, d^+-probability is applicable in not only independent groups, but also in correlated groups. When the two groups are independent, d^+-probability is the probability that a random value from the first group is larger than a random value from the second group. In terms of comparing two therapeutic treatments, d^+-probability is the probability that a patient receiving treatment 1 will improve clinically more than another patient receiving treatment 2, which is equivalent to the well-established probabilistic index $P(X > Y)$ or similar terms [1;29;33;37;115;122;125;134;142;184]. When the two groups are correlated, d^+-probability is the probability that a paired difference is greater than zero. The d^+-probability based on the paired difference in a crossover trial can represent the probability that a specific person randomly chosen from the population improve clinically more on treatment 1 than on treatment 2.

The difference of values between two groups is a contrast variable with coefficients $(1, -1)$. Thus we can apply Theorem 1 from Chapter 7 to derive the following the relationships between SSMD β and d^+-probability by replacing λ and c^+-probability with β and d^+-probability, respectively.

1) If D has normal distribution, d^+-probability $= \Phi(\beta)$ where $\Phi(\cdot)$ is a cumulative distribution function of a standard normal distribution $N(0, 1)$.

2) If D has a unimodal distribution with non-zero finite variance, then

$$\begin{cases} d^+\text{-probability} \geq 1 - \dfrac{4}{9\beta^2}, & \text{for } \beta \geq \sqrt{\dfrac{8}{3}} \\[2ex] d^+\text{-probability} \geq \dfrac{4}{3} - \dfrac{4}{3\beta^2}, & \text{for } 1 \leq \beta \leq \sqrt{\dfrac{8}{3}} \\[2ex] d^+\text{-probability} \leq \dfrac{4}{9\beta^2}, & \text{for } \beta \leq -\sqrt{\dfrac{8}{3}} \\[2ex] d^+\text{-probability} \leq \dfrac{4}{3\beta^2} - \dfrac{1}{3}, & \text{for } -1 \geq \beta \geq -\sqrt{\dfrac{8}{3}} \, . \end{cases}$$

3) If D has a symmetric unimodal distribution with non-zero finite variance, then

$$\begin{cases} d^+\text{-probability} \geq 1 - \dfrac{2}{9\beta^2}, & \text{for } \beta \geq \sqrt{\dfrac{8}{3}} \\[2ex] d^+\text{-probability} \geq \dfrac{7}{6} - \dfrac{2}{3\beta^2}, & \text{for } 1 \leq \beta \leq \sqrt{\dfrac{8}{3}} \\[2ex] d^+\text{-probability} \leq \dfrac{2}{9\beta^2}, & \text{for } \beta \leq -\sqrt{\dfrac{8}{3}} \\[2ex] d^+\text{-probability} \leq \dfrac{2}{3\beta^2} - \dfrac{1}{6}, & \text{for } -1 \geq \beta \geq -\sqrt{\dfrac{8}{3}} \, . \end{cases}$$

Because of the relationship between SSMD and d^+-probability, we can classify the magnitude of difference using SSMD and d^+-probability as shown in Table 7.2 of Chapter 7 by replacing SMCV with SSMD. SSMD, as previously defined, is a population parameter that needs to be estimated from observed samples, as explored in the following section.

8.2 Estimation of SSMD

First, we must consider the estimation of SSMD based on unpaired difference. For convenience, the SSMD based on unpaired difference is called unpaired SSMD. Suppose we have one sample (with sample size n_1, sample mean $\bar{Y}_1$, and sample standard deviation s_1) from group 1 and another independent sample (with n_2, $\bar{Y}_2$, and s_2) from group 2. Let $N = n_1 + n_2$. Then, similarly as in Chapter 7, we can derive the estimation of SSMD as follows.

When the two groups independently have normal distributions, applying Theorem 2 in Chapter 7 with $m = 1$, $a_0 = 0$, $a_1 = 1$, $a_2 = -1$, and $g = 2$, we obtain the following results regarding the estimation of SMCV. When the two groups have unequal variance, the MLE of SSMD (from Formula T2.1 in Theorem 2) is

$$\hat{\beta}_{\text{MLE}} = \frac{\bar{Y}_1 - \bar{Y}_2}{\sqrt{\dfrac{n_1 - 1}{n_1}s_1^2 + \dfrac{n_2 - 1}{n_2}s_2^2}} \tag{8.4}$$

the MM of SSMD (from Formula T2.2) is

$$\hat{\beta}_{\text{MM}} = \frac{\bar{Y}_1 - \bar{Y}_2}{\sqrt{s_1^2 + s_2^2}} \tag{8.5}$$

From Formula T2.3, the asymptotical distribution of $\hat{\beta}_{\text{MLE}}$ is $\hat{\beta}_{\text{MLE}} \sim N(\beta, \sigma_{\hat{\beta}}^2)$ where

$$\hat{\sigma}_{\hat{\beta}}^2 = \frac{\frac{n_1-1}{n_1^2}s_1^2 + \frac{n_2-1}{n_2^2}s_2^2}{\frac{n_1-1}{n_1}s_1^2 + \frac{n_2-1}{n_2}s_2^2} + \left(\frac{(n_1-1)^2}{n_1^3}s_1^4 + \frac{(n_2-1)^2}{n_2^3}s_2^4\right)\frac{(\bar{X}_1-\bar{X}_2)^2}{2\left(\frac{n_1-1}{n_1}s_1^2 + \frac{n_2-1}{n_2}s_2^2\right)^3}.$$

Thus the $1-\alpha$ confidence interval of β is $\hat{\beta} \pm Z_{\frac{\alpha}{2}}\hat{\sigma}_{\hat{\beta}}$ [162]. Z_α is defined such that $\Pr(Z \leq Z_\alpha) = 1-\alpha$ and Z is a standard normal distribution. This is an approximation, especially when the sample size is small.

In a situation with unequal variance but equal sample size r in each group, from Formula T2.4, we have

$$\frac{\bar{Y}_1 - \bar{Y}_2}{\sqrt{\frac{s_1^2 + s_2^2}{r}}} \sim t\left(\nu, \beta\sqrt{r}\right) \text{ approximately, } \nu = (r-1)\frac{\left(s_1^2 + s_2^2\right)^2}{s_1^4 + s_2^4}. \tag{8.6}$$

Thus, regardless of whether the sample size is large or small, one can use the noncentral t-distribution in Formula 8.6 to get an approximate confidence interval, as follows. Let $F_{t(\nu, b\lambda)}(\cdot)$ be the cumulative distribution function of noncentral $t(\nu, b\lambda)$ where $b = \sqrt{r}$ and ν is shown in Formula 8.6. Let T_{obs} be the observed value of T, namely $T_{\text{obs}} = \sqrt{r}(\bar{Y}_1 - \bar{Y}_2)/(\sqrt{s_1^2 + s_2^2})$. Then we can find λ_L and λ_U such that $F_{t(\nu, b\lambda_L)}(T_{\text{obs}}) = 1 - \frac{\alpha}{2}$ and $F_{t(\nu, b\lambda_u)}(T_{\text{obs}}) = \frac{\alpha}{2}$; subsequently (λ_L, λ_U) is approximately a $1-\alpha$ confidence interval of SMCV λ, and an approximate unbiased estimate (from Formula T2.5 in Chapter 7) is

$$\hat{\lambda}_{\text{AUE}} = \sqrt{\frac{2}{\nu}}\frac{\Gamma\left(\frac{\nu}{2}\right)}{\Gamma\left(\frac{\nu-1}{2}\right)}\frac{\bar{Y}_1 - \bar{Y}_2}{\sqrt{s_1^2 + s_2^2}} \tag{8.7}$$

When the two independent groups have normal distributions with equal variance, the UMVUE of unpaired SSMD [161] is and $\hat{\beta}_{UMVUE} = (\bar{Y}_1 - \bar{Y}_2)/\sqrt{\frac{2}{K}\left((n_1-1)s_1^2 + (n_2-1)s_2^2\right)}$, $K = 2 \cdot \left(\Gamma\left(\frac{N-2}{2}\right)/\Gamma\left(\frac{N-3}{2}\right)\right)^2$ when $n_1 \geq 2$, $n_2 \geq 2$. From Figure 8.1, $K \approx N - 3.48$. In primary HTS experiments, $n_1 = 1$ for most investigated siRNAs. s_1^2 does not exist when $n_1 = 1$. In this case, the UMVUE of unpaired SSMD is then $\hat{\beta}_{UMVUE} = (Y_{11} - \bar{Y}_2)/\sqrt{\frac{2}{K}(n_2-1)s_2^2}$, where $K = 2 \cdot \left(\Gamma\left(\frac{n_2-1}{2}\right)/\Gamma\left(\frac{n_2-2}{2}\right)\right)^2 \approx n_2 - 2.48$. If set $(n_1-1)s_1^2 = 0$ when $n_1 = 1$, then for both $n_1 = 1$ and $n_1 \geq 2$ (i.e., for $n_1 \geq 1$),

$$\hat{\beta}_{UMVUE} = \frac{\bar{Y}_1 - \bar{Y}_2}{\sqrt{\frac{2}{K}\left((n_1-1)s_1^2 + (n_2-1)s_2^2\right)}} \tag{8.8}$$

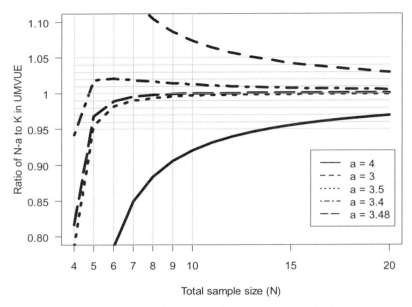

Figure 8.1   Find an approximation in the form of N-a for the constant K in the UMVUE of SSMD.

where $K = 2 \cdot \left(\Gamma(\frac{N-2}{2})/\Gamma(\frac{N-3}{2})\right)^2 \approx N - 3.48$, $v = N - 2$. The MM estimate of unpaired SSMD is

$$\hat{\beta}_{MM} = \frac{\bar{Y}_1 - \bar{Y}_2}{\sqrt{\frac{2}{N-2}\left((n_1 - 1)s_1^2 + (n_2 - 1)s_2^2\right)}} \tag{8.9}$$

$$T = \frac{(\bar{Y}_1 - \bar{Y}_2)\Big/\sqrt{\frac{1}{n_1} + \frac{1}{n_2}}}{\sqrt{\frac{1}{N-2}\left((n_1 - 1)s_1^2 + (n_2 - 1)s_2^2\right)}} \sim \text{noncentral } t\left(N - 2, \frac{\sqrt{2}}{\sqrt{\frac{1}{n_1} + \frac{1}{n_2}}}\beta\right) \tag{8.10}$$

The noncentral t-distribution in Formula 8.10 can be used to derive the confidence interval of SSMD when the two groups have equal variances.

In confirmatory or primary screens with replicates, we are interested in the paired difference between an siRNA and a negative reference in each plate. The SSMD corresponding to paired difference is called *paired SSMD*, as compared with unpaired SSMD. Suppose we observe n pairs of samples, $(Y_{11}, Y_{21}), (Y_{12}, Y_{22}), \ldots, (Y_{1n}, Y_{2n})$ from populations P_1 and P_2, respectively. Let D_j be the difference between the jth pair of samples, namely $D_j = Y_{1j} - Y_{2j}$. Let $\bar{D}$ and s_D be the sample mean and sample standard deviation of D, respectively, namely $\bar{D} = \frac{1}{n}\sum_{j=1}^{n} D_j$ and $s_D^2 = \frac{1}{n-1}\sum_{j=1}^{n}(D_j - \bar{D})^2$. Assume that D is normally distributed, namely $D \sim N(\mu_D, \sigma_D^2)$. Applying Theorem 2 with $g = 1, m = 1, a_0 = 0, a_1 = 1, n_1 = n, \bar{Y}_1 = \bar{D}$, and $s_1^2 = s_D^2$, we have the MM, MLE, and UMVUE of the paired SSMD being (from T2.6–T2.8):

$$\hat{\beta}_{UMVUE} = \frac{\Gamma\left(\dfrac{n-1}{2}\right)}{\Gamma\left(\dfrac{n-2}{2}\right)} \sqrt{\frac{2}{n-1}} \frac{\bar{D}}{s_D} \tag{8.11}$$

$$\hat{\beta}_{MLE} = \sqrt{\frac{n}{n-1}} \frac{\bar{D}}{s_D}, \tag{8.12}$$

$$\hat{\beta}_{MM} = \frac{\bar{D}}{s_D}, \tag{8.13}$$

respectively, and (from T2.9)

$$T = \frac{\sqrt{n}\bar{D}}{s_D} \sim \text{noncentral } t(n-1, \sqrt{n}\beta). \tag{8.14}$$

Given this noncentral distribution T in Formula 8.14, we can obtain the confidence interval of SSMD β. That is, let $F_{t(v, b\beta)}(\cdot)$ be the cumulative distribution function of noncentral $t(v, b\beta)$ (where $v = n-1$ and $b = \sqrt{n}$) and T_{obs} be the observed value of T. Then we can find β_L and β_U such that $F_{t(v, b\beta_L)}(T_{\text{obs}}) = 1 - \frac{\alpha}{2}$ and $F_{t(v, b\beta_u)}(T_{\text{obs}}) = \frac{\alpha}{2}$; subsequently (β_L, β_U) is a $1 - \alpha$ confidence interval of SSMD β. Note that traditional contrast analysis relies on point estimate $\bar{D}$, $1 - \alpha$ confidence interval $\bar{D} \pm T_{n-1, \frac{\alpha}{2}} \frac{s_D}{\sqrt{n}}$ of μ_D, and p-value based on $T = \sqrt{n} \frac{\bar{D}}{s_D} \sim$ central $t(n-1)$ under $H_0 : \mu_D = 0$, which can be applied to the inference of mean difference.

8.3 Comparing SSMD with Standardized Mean Difference and Classical *t*-Statistic

8.3.1 Classical *t*-Test and Standardized Mean Difference

The classical t-test for testing mean difference has been widely used for the comparison of two groups. This t-test is based on the t-statistic that has the following t distributions from T2.11 and T2.13. Under the null hypothesis of $H_0 : \mu_1 = \mu_2$ (or, equivalently, $H_0 : \mu_D = \mu_1 - \mu_2 = 0$),

$$t\text{-statistic} = \begin{cases} \dfrac{\bar{Y}_1 - \bar{Y}_2}{\sqrt{\left(\dfrac{1}{n_1} + \dfrac{1}{n_2}\right)\left(\dfrac{n_1 - 1}{N - 2}s_1^2 + \dfrac{n_2 - 1}{N - 2}s_2^2\right)}} \sim t(N-2), & \text{if equal variance;} \\[3em] \dfrac{\bar{Y}_1 - \bar{Y}_2}{\sqrt{\dfrac{s_1^2}{n_1} + \dfrac{s_2^2}{n_2}}} \sim t(v) \text{ approximately,} & \text{if not equal variance;} \end{cases}$$

$$\tag{8.15}$$

where ν is the integer part of $\dfrac{\left(s_1^2/n_1+s_2^2/n_2\right)^2}{\left(s_1^2/n_1\right)^2\big/(n_1-1)+\left(s_2^2/n_2\right)^2\big/(n_2-1)}$ according to the Satterth-waite option. The z-score method is based on normal approximation. When n_1 and n_2 are both large,

$$z\text{-score} = \frac{\bar{Y}_1 - \bar{Y}_2}{\sqrt{\dfrac{s_1^2}{n_1} + \dfrac{s_2^2}{n_2}}} \sim N(0, 1) \tag{8.16}$$

under the null hypothesis. In the situation with equal sample sizes in the two samples, namely $n_1 = n_2 = \frac{N}{2}$, Formula 8.15 becomes

$$t\text{-statistic} = \frac{\bar{Y}_1 - \bar{Y}_2}{\sqrt{\dfrac{2}{N}(s_1^2 + s_2^2)}} \sim t(\nu) \text{ approximately} \tag{8.17}$$

where ν is $N-2$ if equal variance and is the integer part of $\frac{N-2}{2}\frac{(s_1^2+s_2^2)^2}{s_1^4+s_2^4}$ if not equal variance, and Formula 8.16 becomes

$$z\text{-score} = \frac{\bar{Y}_1 - \bar{Y}_2}{\sqrt{\dfrac{2}{N}(s_1^2 + s_2^2)}} \sim N(0, 1) \tag{8.18}$$

Formulas 8.15 and 8.16, and especially Formulas 8.17 and 8.18, clearly indicate that sample size strongly affects the t-statistic and z-score. That is, sample size and magnitude of difference are indistinguishable in the statistical significance (i.e., p-value from a t-test or z-score). As a result, we may obtain a significant result in an experiment with a very small magnitude of difference but a large sample size, whereas we may obtain a nonsignificant result in an experiment with a large magnitude of difference but a small sample size. Thus the p-values are hardly comparable from various experiments with different sample sizes. Largely due to this feature in the statistical significance, the use of t-statistic for the comparison of two groups has been intensively criticized in medical and social sciences [4;26;34;55;75;76;84;107;114;141]. Thus effect size has been proposed for group comparisons [32;53;153].

There are various types of effect sizes. One popular type is standardized mean difference, namely $d = \frac{\mu_1-\mu_2}{\sigma}$ (cf. [33;67]). The widely used estimate is

$$\text{Cohen's } d = \frac{\bar{Y}_1 - \bar{Y}_2}{\sqrt{\dfrac{1}{N-2}\left((n_1-1)s_1^2 + (n_2-1)s_2^2\right)}} \tag{8.19}$$

When there are treatment and control groups, Glass [53] suggested using sample standard deviation in the control group. That is, assuming the second group is the control group,

$$\text{Glass's } \delta = \frac{\bar{Y}_1 - \bar{Y}_2}{s_2} \tag{8.20}$$

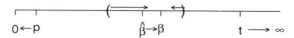

Figure 8.2   The moving trends of SSMD estimate $\hat{\beta}$, the bounds of its confidence interval (CI), *t*-statistic, and *p*-value as sample size *N* increases. The parentheses denote the CI bounds. β denotes the true value of SSMD. As $N \to \infty$, *t*-statistic approaches ∞ or $-\infty$ and *p*-value approaches zero; however, the estimate $\hat{\beta}$ and the CI lower and upper bounds all approach β.

The two estimates in Equations 8.19 and 8.20 are all based on the assumption of equal variances in the two groups. Standardized mean difference has rarely been investigated in a situation in which the two groups have unequal variances, although Cohen [32] proposes the definition of $d = (\mu_1 - \mu_2)/\sqrt{(\sigma_1^2 + \sigma_2^2)/2}$ for standardized mean difference in this situation.

8.3.2 Comparing SSMD with Classical *t*-Test

From Formulas 8.17, 8.18, 8.8 and 8.9, in the situations with equal sample sizes, the relationship between *t*-statistic (or *z*-score) and SSMD estimate can be as simple as

$$t\text{-statistic} = \sqrt{\frac{N}{2}}\hat{\beta}_{MM} = \sqrt{\frac{N}{2}}\sqrt{\frac{N-2}{K}}\hat{\beta}_{UMVUE}, \tag{8.21}$$

which indicates that *t*-statistic and associated *p*-value are affected not only by the magnitude of difference, but also by sample size. In general, when there is even any tiny mean difference, larger sample size (N) leads to large absolute value of *t*-statistic and smaller *p*-value. In fact, as $N \to \infty$, *t*-statistic value approaches ∞ or $-\infty$ and *p*-value approaches to 0 (Figure 8.2). By contrast, as sample size increases, the SSMD estimate approaches the true SSMD value in probability (Figure 8.2). Sample size can only impact the precision of $\hat{\beta}$ in representing β. Thus we can still have the benefit of increasing sample size in an experiment: increasing the precision of SSMD estimation. When there are unequal sample sizes in the two groups, the relationship between *t*-statistic and SSMD estimate can be complicated; however, the moving trends of SSMD estimate, *t*-statistic, and *p*-value are still the same, as illustrated in Figure 8.2.

In traditional methods for testing mean difference μ_D using *t*-statistic or *z*-score, one may also use both the point estimate and confidence interval of μ_D. However, they will lead to the same misleading results as the use of *p*-value for testing mean difference. This is because as $N \to \infty$, the point estimate and the upper and lower bounds of the confidence interval all approach the population value of μ_D (Figure 8.2), and it is well known that the mean difference μ_D alone cannot represent the magnitude of difference between two groups.

Here I use a simple simulated example to illustrate the issues with the use of *p*-value for testing mean difference and the point estimate and confidence interval for mean difference. In this example, two studies are conducted to investigate two treatments for reducing blood pressure. Suppose the blood pressures in people

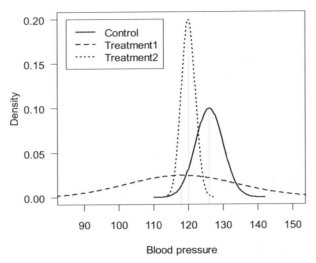

Figure 8.3   Population distributions of blood pressures in patients who received two treatments, treatment 1 and treatment 2, and patients who received no treatment (i.e., control).

without treatment (i.e., controls), with treatment 1, and with treatment 2 have normal distributions $N(126, 4^2)$, $N(118, 16^2)$, and $N(120, 2^2)$, respectively. The interest is in the difference between blood pressure with a treatment and without a treatment. In study 1, there are 800 people without treatment and 800 with treatment 1; in study 2, there are 100 people without treatment and 100 with treatment 2.

Given the two simulated studies, the p-value for testing no mean difference between treatment 1 and control is 6×10^{-42}; the mean difference between treatment 1 and control is estimated to be -8.10, and its 95% confidence interval is $(-9.22, -6.99)$. By contrast, the p-value for testing no mean difference between treatment 2 and control is 2×10^{-21}; the mean difference between treatment 2 and control is estimated to be -5.88 and its 95% confidence interval is $(-6.88, -4.88)$. If one judges based on p-value, the estimate of mean difference, and its confidence interval, one would easily conclude that treatment 1 is better than treatment 2 in reducing blood pressure. This is because treatment 1 has a much smaller p-value and a larger amount of reduction in blood pressure on average and because the upper bound of the confidence interval for treatment 1 is smaller than the lower bound of the confidence interval for treatment 2. However, the comparison of distributions clearly reveals that treatment 2 is better than treatment 1, even though treatment 2 has a smaller amount of reduction on average (Figure 8.3).

Using Formula 8.4, the MLE estimate and 95% confidence interval of SSMD are -0.506 and $(-0.579, -0.433)$, respectively, for treatment 1 and -1.308 and $(-1.582, -1.0346)$, respectively, for treatment 2, which correctly indicate that treatment 2 has a large effect whereas treatment 1 has only a medium effect in reducing blood pressure. The use of d^+-probability also reaches a similar result. For treatment 1, the d^+-probability is 0.69, which means that the chance that patients receiving treatment

1 have lower blood pressure than those who did not receive a treatment is 69%. In contrast, for treatment 2, the d^+-probability is 0.90, which means that the chance that patients receiving treatment 2 have lower blood pressure than those who did not receive a treatment is 90%. Thus the d^+-probability indicates that there is a much larger chance of reducing blood pressure for a person taking treatment 2 than another person taking treatment 1. All these indicate that SSMD and d^+-probability provide a potentially better alternative to evaluate treatment effect than p-value and confidence interval of mean difference in traditional approaches.

In many studies, the real interest is the magnitude of difference, not sample size. In those studies, the p-value is actually not comparable across experiments with different sample sizes. Therefore, applying the same cutoff criterion such as a p-value of 0.05 cannot address the question of interest in many studies. In drug development, a p-value of 0.05 or 0.01 has been used to indicate safety and efficacy. We must be aware that even if there is only minor toxicity, a large sample size can lead to a very small p-value; on the other hand, mid-grade or major toxicity may not result in a small p-value if the sample size is small. The p-value or statistical significance is therefore not a good indication for drug safety and efficacy.

8.3.3 Comparing SSMD with Standardized Mean Difference

In a situation in which two groups do not have correlation, the relationship between Cohen's d and SSMD is

$$d = \sqrt{2}\beta. \tag{8.22}$$

Thus far, no research has been conducted to investigate standardized mean difference in a situation in which two groups have correlations. The definition of SSMD is applicable not only in independent groups, but also in correlated groups. For two correlated groups, we can either estimate the covariance σ_{12} between two groups or use paired sample to estimate SSMD directly on the basis of paired differences. In addition, SSMD can be extended to comparison involving more than two groups [163;167;170]. Therefore, SSMD has a broader utility than Cohen's d in two-group comparisons.

SSMD has two clear and meaningful interpretations when it is used to assess the magnitude of difference between two groups. The first interpretation is that it is the ratio of mean to standard deviation of a random variable representing the difference between two groups, and the second interpretation is that it reflects the probability that a random value from the first group is larger than a random value from the second group, namely d^+-probability [161;162;165]. When the data are normally distributed in both groups, d^+-probability $= \Phi(\beta)$, where $\Phi(\cdot)$ is the cumulative density function of the standard normal distribution. When the data are not normally distributed, there is still a relationship between d^+-probability and β [166]. Because of clear and meaningful interpretations of SSMD, we can construct meaningful and interpretable SSMD-based criteria for classifying the magnitude of the difference between two groups, namely $|\beta| \geq 1.645$ for *extra large*, $1.645 > |\beta| \geq 1$ for *large*, $1 > |\beta| \geq 0.5$ for *medium large*, $0.5 > |\beta| > 0.25$ for *medium*, and $|\beta| \leq 0.25$ for *small* under normality assumption [167]. That is, an extra-large effect for a comparison indicates that a ratio of mean to variability of a contrast variable representing the comparison is at least 1.645, a large effect indicates a mean-to-variability ratio between 1 and 1.645, a medium-large effect indicates a mean-to-variability ratio between 0.5 and 1, a medium effect indicates a mean-to-variability ratio between 0.25 and 0.5, and a small effect indicates a mean-to-variability ratio between 0 and 0.25. Under normality assumption, these SSMD-based criteria can be interpreted using probability as follows. An extra-large effect means that the probability of a value from the first group being greater than a value from the second group is greater than 0.95 in the positive direction of the difference and is less than 0.05 in the negative direction. A large effect means that this probability (i.e., d^+-probability) is between 0.84 and 0.95 in the positive direction and between 0.05 and 0.16 in the negative direction. A medium-large effect means that this probability is between 0.7 and 0.84 in the positive direction and is between 0.16 and 0.3 in the negative direction. A medium effect means that this probability is between 0.6 and 0.7 in the positive direction and is between 0.16 and 0.3 in the negative direction. Finally, a small effect means that this probability is between 0.5 and 0.6 in the positive direction and is between 0.40 and 0.50 in the negative direction.

Given the relationship between SSMD and standardized mean difference, the original and probability meanings of SSMD also give clear interpretations to Cohen's and McLean's criteria [33;104]. Cohen's $d = 0.20, 0.50$, and 0.80 correspond to SSMD $= 0.1414, 0.3536$, and 0.5656, respectively, and d^+-probability $= 0.556, 0.638$, and 0.714, respectively. Hence Cohen's small, medium, and large effects have a mean-to-variability ratio of 14.14%, 35.36%, and 56.56%, respectively, and a d^+-probability of 0.556, 0.638, and 0.714, respectively, under normality. In other words, if one randomly draws one value from each of two groups, Cohen's small, medium, and large effects indicate that the chance that the value from the first group is greater than that from the second group is 0.556, 0.638, and 0.714, respectively, in the positive direction. Cohen's $d = (0, 0.50), (0.50, 1.00)$, and $(1.00, \infty)$ correspond to SSMD $= (0, 0.3536), (0.3536, 0.7071)$, and $(0.7071, \infty)$, respectively, and d^+-probability $= (0.50, 0.64), (0.64, 0.76)$, and $(0.76, 1)$, respectively. That is, McLean's small, moderate, and large effects have a mean-to-variability ratio between 0% and 35.36%, between 35.36% and 70.71%, and greater than 70.71%, respectively. Given the d^+-probability, if one randomly draws one value from each of two groups, McLean's small, moderate, and large effects indicate that the chance that the value from the first group is greater than that from the second group is between 0.5 and 0.64, between 0.64 and 0.76, and greater than 0.714, respectively, in the positive direction. Therefore, SSMD criterion gives interpretations to Cohen's and McLean's criteria from both strength and probability perspectives. The small, medium, and large effects based on Cohen's criteria are, respectively, small, medium, and medium large effects based on SSMD. McLean's small and medium effects are roughly equivalent to SSMD's small and medium effects, respectively. McLean's large effects contain SSMD's medium large, large, and extra large effects.

In summary, when two groups are independent, there is a simple relationship between Cohen's d and SSMD. However, Cohen's d cannot be applied to a situation in which two groups are not independent, whereas SSMD can readily be applied to measure the magnitude of difference when groups are correlated. Therefore, SSMD has a broader utility than standardized mean difference. In addition, SSMD has clear and meaningful interpretations. These interpretations also provide a base for interpreting Cohen's criterion. Therefore, SSMD not only has a broader utility than Cohen's d in measuring effect sizes, but also offers a good interpretation to Cohen's and McLean's criteria for classifying effect size.

8.4 SSMD-Based Ranking Methods for Hit Selection in Genome-Scale RNAi Screens

There are two main strategies for selecting hits with large effects. One is to use certain metric(s) to rank the siRNAs by their effects and then to select the largest number of potent siRNAs that is practical for confirmation and validation assays. The other strategy is to test whether an siRNA has strong enough effects to reach a pre-set specified effect in which we need to control the FNRs and/or FPRs. In the first strategy, traditionally, two types of measures are used to rank siRNA effects. One is mean difference, along with its variants such as signal-to-noise ratio and percent inhibition, and the other is p-value from either z-score method or t-test of testing mean difference. The first measure cannot represent the magnitude of difference because it does not effectively capture data variability [159;162;165]. When statistical significance is used, the p-value comes from testing the hypothesis of no mean difference between two groups. It addresses the question of whether an siRNA has exactly the same effect as the negative reference based on the sample observation. It is not designed to measure how large the magnitude of difference is [34;84]. Thus an siRNA effect that results in a low p-value may not cause a robust enough effect on the assay to indicate any meaningful biological association. Therefore, neither mean difference nor p-value can represent the magnitude of difference.

Zhang [162] proposed SSMD as a better metric to measure the magnitude of difference. As described in the previous sections, unlike mean difference and percent inhibition, SSMD is robust to both measurement unit and strength of positive controls; it takes into account data variability in both compared groups and has a probability interpretation [161;162]. Compared with p-value, SSMD directly measures how large the magnitude of difference is [165]. Therefore, SSMD can serve as an effective metric to rank siRNA effects. Consequently, when using the ranking strategy for selecting hits, we can select the manageable number of siRNAs with the largest SSMD value in the up-regulated direction and/or siRNAs with the smallest

SSMD value in the down-regulated direction. When sample size is large, we can further use the classifying criteria listed in Table 7.2 of Chapter 7 to group siRNAs on the basis of their effect sizes. When the sample size is small, we need to control FPRs and FNRs, as described in the following section.

8.5 SSMD-Based FPR, FNR, and Power

8.5.1 FPR, FNR, and Power in RNAi Screens

In genome-scale RNAi screen experiments, the primary interests are (i) the assessment of the magnitude of impact on a biological response related to the knockdown of a gene, and (ii) the selection of siRNAs with large effects on the biological response of interest. The key is to search an analytic metric to effectively quantify knockdown effect and then to construct a selection criterion based on this metric to control FPRs and FNRs.

It is well known that cells are controlled by dynamic actions of thousands of genes that are related through a complex interaction. Because of the existence of gene networks, the knockdown of any gene by its corresponding siRNA(s) may affect other genes, even though the size of effect may differ. Thus, in a broad sense, an siRNA rarely has exactly no impact on a measured biochemical response, although most siRNAs have small effects in a genome-scale RNAi screen. Therefore, there are two major concerns for hit selection in RNAi HTS experiments. First, we do not want the siRNAs with large effects to be selected as non-hits. Second, we do not want the siRNAs with small effects to be selected as hits. The rate in which the siRNAs with small effects are selected as hits is the FPR, and the rate in which the siRNAs with large effects are not selected as hits is the FNR (see also Chapter 5). To control FPR and FNR, we first need a metric that effectively measures the size of siRNA effects. Mean difference, fold change, percent viability, p-value, and other similar measures cannot effectively measure the size of siRNA effects (represented by the magnitude of difference between an investigated siRNA and a negative reference), whereas SSMD can [161;162;165;166;175].

Because SSMD can effectively measure the size of siRNA effects and its value is comparable across experiments[161;162;165;181], we can use a value β_1 of SSMD to indicate large effects and another value β_2 of SSMD to indicate small effects. Given the original and probability meanings of SSMD, the values of interest for β_2 may be 0, 0.25, or 0.5 and those for β_1 may be 1.645, 2, 3, or 5 for SSMD across various HTS experiments [161;167]. For selecting hits with positive effects, we may use the decision rule of declaring an siRNA as a hit if it has $\hat{\beta} \geq \beta^*$ (where $\hat{\beta}$ is estimated SSMD value and β^* is a cutoff) and as a non-hit otherwise. In this selection process, FPR is the probability of selecting an siRNA with $\beta \leq \beta_2$ as a hit, namely Pr(Declare a hit given $\beta \leq \beta_2$). Clearly, in this selection process, the smaller the β value, the smaller the FPR; consequently, given $\beta \leq \beta_2$, the maximum of FPR is achieved when $\beta = \beta_2$. This maximal FPR is the FPL. In other

A: Classical Hypothesis Test

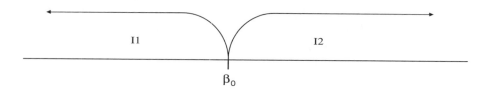

B: SSMD-based Error Control Method

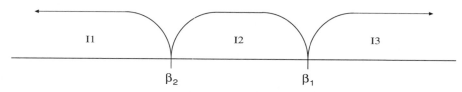

Figure 8.4   SSMD intervals for classical hypothesis testing and error control. (A) Two intervals used in a typical hypothesis test in which β_0 is usually fixed. (B) Three intervals for the flexible and balanced control of FPR and FNR in which β_1 and β_2 are not fixed to a single value and multiple values of β_1 and β_2 are considered simultaneously, as in an SSMD-based method.

words, the upper limit of FPR given $\beta \leq \beta_2$ in a selection process is FPL[161]. Similarly, FNR is the probability of not selecting an siRNA with $\beta \geq \beta_1$ as a hit, namely Pr(Not declare a hit given $\beta \geq \beta_1$). The upper limit of FNR given $\beta \geq \beta_1$ in this selection process is the FNL[161], which is achieved when $\beta = \beta_1$. Note, FPL becomes classical false positive level when $\beta_1 = \beta_2$. Because the true value of β is unknown for each siRNA, what we can control are FPL and FNL. FPL and FNL are also what we really want to control because if we control FPL $= \alpha_1$ for $\beta \leq \beta_2$ and control FNL $= \alpha_2$ for $\beta \geq \beta_1$, then the FPRs for all siRNAs with $\beta \leq \beta_2$ are controlled to be no more than α_1 and the FNRs for all siRNAs with $\beta \geq \beta_1$ are controlled to be no more than α_2.

8.5.2 Classical Hypothesis Test and SSMD-Based Error Control Method

FNR and FPR may also be linked to type I and II error rates under certain conditions [178]. To understand the differences and links between the concepts of FNR and FPR and the concepts of type I and II error rates, we need to explore the difference between classical hypothesis tests and the SSMD-based error control methods. Their major differences are briefly described as follows. First, the SSMD-based method focuses on the control of error rates related to intervals I1 and I3 in Figure 8.4B, whereas classical hypothesis tests aim at the control of error rates in intervals I1 and I2 in Figure 8.4A. Second, for the error control in the SSMD-based method, β_1 and β_2 are not fixed to a single value; instead, multiple meaningful values β_1 and β_2 for the parameter β are considered simultaneously. Third, the classical z-score method or t-test for hit selection usually aims at testing mean difference, fold change, percent viability, and so forth, whereas SSMD-based error control methods aim at SSMD. More details about the difference between the classical hypothesis test and the SSMD-based error control method are described below.

In a classical hypothesis test, there are two hypotheses: (i) a null hypothesis H_0 and (ii) an alternative hypothesis H_a. The alternative hypothesis is the complement of the null hypothesis. For example, the hypothesis testing for a parameter β may be constructed in the following three ways: (i) a two-sided test $H_0 : \beta = \beta_0$ vs. $H_a : \beta \neq \beta_0$, (ii) a one-sided test $H_0 : \beta \leq \beta_0$ vs. $H_a : \beta > \beta_0$, and (iii) a one-sided test $H_0 : \beta \geq \beta_0$ vs. $H_a : \beta < \beta_0$. These hypothesis tests focus on testing a fixed value c for β, as illustrated in Figure 8.4A. The type I error rate is the probability of rejecting H_0 when H_0 is correct (or given the true value in H_0), and the type II error rate is the probability of not rejecting H_0 when H_a is correct. The power of a test is the probability of rejecting H_0 when H_a is correct. Thus power $= 1 -$ type II error rate. The maximum of the type I error rate given any true value in H_0 is called the type I error level; similarly, the maximum of type II error rate given any true value in H_a is called the type II error level.

Because H_a is the complement of H_0 in a classical hypothesis test, the control of the type I error level in a specified decision rule leads to the determination of the type II error level. For example, suppose an estimate $\hat{\beta}$ has a normal distribution. For a one-sided test $H_0 : \beta \leq \beta_0$ vs. $H_a : \beta > \beta_0$, a decision is as follows: reject H_0 in favor of H_a if $\hat{\beta} > \beta^*$ and its corresponding type I error rate is $\Pr(\hat{\beta} > \beta^* | H_0)$. If the type I error level is α, then $\beta^* = c + Z_\alpha \hat{\sigma}_{\hat{\beta}}$, where Z_α is the upper αth quantile of the standard normal distribution (i.e., $\Phi(Z_\alpha) = 1 - \alpha$ where $\Phi(\cdot)$ is the cumulative distribution function of the standard normal distribution) and $\hat{\sigma}_{\hat{\beta}}$ is the estimated standard deviation of $\hat{\beta}$. The corresponding type II error rate is $\Pr(\hat{\beta} \leq \beta_0 + Z_\alpha \hat{\sigma}_{\hat{\beta}} | H_a) = \Pr(\hat{\beta} \leq \beta_0 + Z_\alpha \hat{\sigma}_{\hat{\beta}} | \beta > c) \leq \Phi(Z_\alpha) = 1 - \alpha$. Thus, once the type I error level is set to be α, then the type II error level is determined to be $1 - \alpha$. That is why it is impossible to control both type I and II error levels simultaneously in a classical test for a given data set.

By contrast, for the error control in the SSMD-based method, β_1 and β_2 are not fixed to a single value (Figure 8.4B); instead, multiple meaningful values β_1 and β_2 for the parameter β are considered simultaneously. Let us focus on the situation where $\beta_1 > \beta_2 \geq 0$. The interest is to determine a decision rule in selecting hits with large positive effects (namely, to declare a hit if $\hat{\beta} \geq \beta^*$) so that both

$$\Pr(\text{Declare a hit} \mid \beta \leq \beta_2) \tag{8.23}$$

and

$$\Pr(\text{Not declare a hit} \mid \beta \geq \beta_1) \tag{8.24}$$

are under control. In other words, the core idea is to select a set of values for $(\beta^*, \beta_1, \beta_2)$ such that both $\Pr(\hat{\beta} \geq \beta^* | \beta \leq \beta_2)$ and $\Pr(\hat{\beta} < \beta^* | \beta \geq \beta_1)$ are reasonably low. The rationale behind this approach is the consideration of two major concerns in the process of selecting the siRNAs with a large effect in RNAi HTS experiments. First, we do not want the siRNAs with large effects to be selected as non-hits. Second, we do not want the siRNAs with small effects to be selected as hits. The probabilities $\Pr(\text{Declare a hit}|\beta \leq \beta_2)$ and $\Pr(\text{Not declare a hit}|\beta \geq \beta_1)$ are FPR and FNR, respectively; their upper limits are FPL and FNL, respectively [161]. When $\beta_1 = \beta_2$, FPR becomes classical false positive rate and FPL becomes classical false positive level.

8.5.3 Controlling FPR and FNR in Genome-Scale RNAi Screens

To use the SSMD-based method for selecting hits in the direction of positive values in HTS assays, we need to search for a threshold β^* for the estimated SSMD $\hat{\beta}$ to maintain a balanced control of both FPL and FNL when we use the decision rule of declaring an siRNA as a hit if it has $\hat{\beta} \geq \beta^*$ and as a non-hit otherwise. The search for the decision rule relies on the distribution of SSMD estimates. When sample size is large, the estimate of SSMD has approximately a normal distribution [162]. Given β^*, β_1, and β_2, we have

$$\text{FPL} = 1 - \Phi\left(\frac{\beta^* - \beta_2}{\hat{\sigma}_{\hat{\beta}}}\right) \tag{8.25}$$

$$\text{FNL} = \Phi\left(\frac{\beta^* - \beta_1}{\hat{\sigma}_{\hat{\beta}}}\right) \tag{8.26}$$

where $\hat{\sigma}_{\hat{\beta}}$ can be calculated using the formulas provided in Zhang [162], and $\Phi(\cdot)$ is the cumulative distribution function of the standard normal distribution. In a primary HTS experiment with no replicates for each siRNA, one approximated value for $\hat{\sigma}_{\hat{\beta}}$ is $\sqrt{\frac{1}{2} + \frac{1}{4n_2}\frac{(\bar{X}_1 - \bar{X}_2)^2}{s_2^2}}$, or simply $\sqrt{\frac{1}{2}}$, when the number of sample wells in a plate is large (e.g., $n_2 \geq 100$). When sample size is small, normal approximation cannot work effectively. However, we can use noncentral t-distribution to calculate FPL and FNL, as described in Zhang [165]. Given β^*, β_1, and β_2, we have

$$\text{FPL} = 1 - F_{t(\nu,b\beta_2)}\left(\frac{\beta^*}{k}\right) \tag{8.27}$$

$$\text{FNL} = F_{t(\nu,b\beta_1)}\left(\frac{\beta^*}{k}\right) \tag{8.28}$$

where $t(\nu, b\beta)$ is a noncentral t-distribution with ν degrees of freedom and noncentral parameter $b\beta$ and $F_{t(\nu,b\beta)}(\cdot)$ is the cumulative distribution function of $t(\nu, b\beta)$. The values of k, ν and b are different for unpaired and paired differences as described next. For an unpaired difference mainly in a primary screen, we can use Formula 8.8 or 8.9 to calculate the estimated values of SSMD. Based on Formula 8.10, $T = (\bar{Y}_1 - \bar{Y}_2)/\left(\sqrt{((n_1 - 1)s_1^2 + (n_2 - 1)s_2^2)/(N - 2)} \cdot \sqrt{\frac{1}{n_1} + \frac{1}{n_2}}\right) \sim$ noncentral $t(\nu, b\beta)$, where $\nu = N - 2$, $b = \sqrt{2}/\sqrt{\frac{1}{n_1} + \frac{1}{n_2}}$; $k = \sqrt{\frac{K}{2(N-2)}\left(\frac{1}{n_1} + \frac{1}{n_2}\right)}$ when using UMVUE of SSMD and $k = \sqrt{\frac{1}{2}\left(\frac{1}{n_1} + \frac{1}{n_2}\right)} = \frac{1}{b}$ when using MM estimate of SSMD. For a paired difference in a confirmatory or primary screen with replicates, we can use Formula 8.11 to calculate the estimated values of SSMD. Based on Formula 8.14, $T = \sqrt{n}\bar{D}/s_D \sim$ noncentral $t(\nu, b\beta)$, where $\nu = n - 1$, $b = \sqrt{n}$; $k = \Gamma(\frac{n-1}{2})/\Gamma(\frac{n-2}{2})\sqrt{\frac{2}{n(n-1)}}$ when using UMVUE of SSMD and $k = \sqrt{\frac{1}{n}} = \frac{1}{b}$ when using MM estimate of SSMD.

In the SSMD-based method for maintaining a balanced control of both FPR and FNR, β_1 and β_2 are not fixed to a single value. To search a decision rule for hit selection, one approach is, given a critical value, to calculate FPLs with respect to certain meaningful β_2's using Formula 8.25 or 8.27 and FNLs with respect to certain meaningful β_1's using Formula 8.26 or 8.28; then choose a critical value that researches acceptable FPL and FNL. There are two additional approaches: (i) pre-set β_1 and FNL and then search β^*, β_2, and FPL based on the relationships among β^*, β_1, β_2, FNL, and FPL; (ii) pre-set β_2 and FPL and then search β^*, β_1, and FNL. For example, under the assumption of normality, given β_1 and FNL, we can search β^*, β_2, and FPL using

$$\beta^* = \beta_1 - Z_{\text{FNL}}\hat{\sigma}_{\hat{\beta}} \tag{8.29}$$

$$\text{FPL} = \Phi\left(Z_{\text{FNL}} - \frac{\beta_1 - \beta_2}{\hat{\sigma}_{\hat{\beta}}}\right). \tag{8.30}$$

On the other hand, given β_2 and FPL, we can search β^*, β_1, and FNL using

$$\beta^* = \beta_2 + Z_{\text{FPL}}\hat{\sigma}_{\hat{\beta}} \tag{8.31}$$

$$\text{FNL} = \Phi\left(Z_{\text{FPL}} + \frac{\beta_2 - \beta_1}{\hat{\sigma}_{\hat{\beta}}}\right). \tag{8.32}$$

The reason why we can let the search begin with either β_1 or β_2 is because the intervals I1 and I3 (specified by β_1 and β_2, respectively) shown in Figure 8.4 are treated equally in the SSMD-based method.

Similar to Formulas 8.29, 8.30, 8.31, and 8.32, the following formulas can be obtained when sample size is small, such as in a confirmatory or primary screen with replicates.

$$\beta^* = k\, Q_{t(\nu, b\beta_1)}(\text{FNL}) \tag{8.33}$$

$$\text{FPL} = 1 - F_{t(\nu, b\beta_2)}\left(Q_{t(\nu, b\beta_1)}(\text{FNL})\right) \tag{8.34}$$

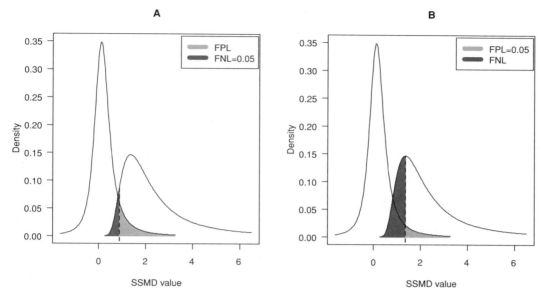

Figure 8.5 False-positive level (FPL) and false-negative level (FNL) displayed in two cases: one controlling FNL = 0.05 with respect to $\beta_1 = 3$ (A) and the other controlling FPL = 0.05 with respect to $\beta_2 = 0.25$ (B).

and

$$\beta^* = k Q_{t(v,b\beta_2)}(1 - \text{FPL}) \tag{8.35}$$

$$\text{FNL} = F_{t(v,b\beta_1)}\left(Q_{t(v,b\beta_2)}(1 - \text{FPL})\right) \tag{8.36}$$

where $F_{t(v,b\beta)}(\cdot)$ and $Q_{t(v,b\beta)}(\alpha)$ are the cumulative distribution function and the α quantile of $t(v, b\beta)$. Figure 8.5A shows the use of Formulas 8.33 and 8.34 to search β^*, β_2, and FPL given β_1 and its corresponding FNL being 0.05 in a screen with triplicates for each investigated siRNAs or pools. Figure 8.5B shows the use of Formulas 8.35 and 8.36 to search β^*, β_1, and FNL given β_2 and its corresponding FPL being 0.05 in a screen with triplicates for each investigated siRNAs or pools.

One can interpret FNR as type I error rate and FPL as type II error or $1 -$ power(β_2) based on the hypotheses $H_0 : \beta \geq \beta_1$ vs. $H_a : \beta < \beta_1$. With regard to these hypotheses, rejecting the null hypothesis means not declaring a hit. Thus type I error is $\Pr(\text{Not declare a hit} \mid H_0 : \beta \geq \beta_1)$ and the power with respect to β_2 is $\Pr(\text{Not declare a hit} \mid \beta = \beta_2 < \beta_1)$. Clearly, FNR = type I error and FPL = $1 -$ power(β_2). This interpretation comes from the approach of using Formulas 8.29 and 8.30 to search a set of values for (β^*, β_1, β_2, FNL, and FPL) for maintaining a flexible and balanced control of both FPR and FNR. And it works only when β_1 is fixed. On the other hand, if we use Formulas 8.31 and 8.32, we can similarly derive another interpretation for FPR and FNR, namely interpreting FPR as type I error rate and FNR as type II error rate when β_2 is fixed. The second interpretation matches with the terms of type I and type II errors in classical z-score method and t-test for hit selection better than the first interpretation. The reason is as follows.

Considering the one-sided z-score method for selecting hits with large positive effects, the hypotheses are $H_0 : \beta \leq 0$ vs. $H_a : \beta > 0$. Suppose we set $\beta_2 = 0$; then the hypotheses become $H_0 : \beta \leq \beta_2$ vs. $H_a : \beta > \beta_2$. Assuming $\hat{\beta}$ approximately has a normal distribution with mean β and variance $\hat{\sigma}_{\hat{\beta}}^2$, the corresponding type I error rate is

$$\Pr(\text{declare a hit} \mid H_0) = \Pr(\hat{\beta} \geq \beta^* \mid \beta \leq \beta_2) \leq 1 - \Phi\left(\frac{\beta^* - \beta_2}{\hat{\sigma}_{\hat{\beta}}}\right) \tag{8.37}$$

If the type I error level is set to be α, we have

$$\beta^* = \beta_2 + Z_\alpha \hat{\sigma}_{\hat{\beta}} \tag{8.38}$$

and the type II error level is $1 - \alpha$. The type I error rate in Formula 8.37 has the same form as the FPR in Formula 8.23 instead of the FNR in Formula 8.24, and the critical value in Formula 8.38 has the same form as that in Formula 8.31 instead of that in Formula 8.29.

8.5.4 p-Value and p^*-Value in RNAi Screens

When selecting up-regulated hits (i.e., siRNAs with large positive value) for an siRNA with an observed value β_{obs} of SSMD, given the true value of SSMD no more than a small value β_2, p-value is defined as the maximum probability of selecting this siRNA as a hit if we use the following selection criterion: any siRNA is selected as a hit if it has the estimated SSMD value no less than β_{obs} and as a non-hit otherwise. That is, p-value is the maximum of $\Pr(\hat{\beta} \geq \beta_{\text{obs}} \mid \beta \leq \beta_2)$, namely p-value $= \Pr(\hat{\beta} \geq \beta_{\text{obs}} \mid \beta = \beta_2)$. Based on the noncentral t-distribution of T in Formula 8.10 or 8.14,

$$p\text{-value} = 1 - F_{t(\nu, b\beta_2)}\left(\frac{\beta_{\text{obs}}}{k}\right) \tag{8.39}$$

where $t(\nu, b\beta)$, ν, b, k and $F_{t(\nu, b\beta)}(\cdot)$ are as in Formula 8.28.

The preceding p-value corresponds to FPR with respect to β_2. In parallel, for convenience, we can define p^*-value as the maximum of $\Pr(\hat{\beta} < \beta_{\text{obs}} \mid \beta \geq \beta_1)$, namely p^*-value $= \Pr(\hat{\beta} < \beta_{\text{obs}} \mid \beta = \beta_1)$, which corresponds to FNR with respect to β_1 [169]. Based on the noncentral t-distribution of T in Formulas 8.10 or 8.14,

$$p^*\text{-value} = F_{t(\nu, b\beta_1)}\left(\frac{\beta_{\text{obs}}}{k}\right). \tag{8.40}$$

The values of β_2 are 0 or 0.25, and the values of β_1 are 3 or 5.

Similarly, when selecting down-regulated hits (i.e., siRNAs with small negative value), for an siRNA with an observed value β_{obs} of SSMD, the p-value with respect to β_2 is the maximum of $\Pr(\hat{\beta} \le \beta_{obs}|\beta \ge \beta_2)$, namely p-value $= \Pr(\hat{\beta} \le \beta_{obs}|\beta = \beta_2)$; thus

$$p\text{-value} = F_{t(v,b\beta_2)}\left(\frac{\beta_{obs}}{k}\right) \tag{8.41}$$

And the p*-value w.r.t. β_1 is, that is,

$$p^*\text{-value} = 1 - F_{t(v,b\beta_1)}\left(\frac{\beta_{obs}}{k}\right). \tag{8.42}$$

The values of β_2 may be 0 or –0.25 and the values of β_1 may be –3 or –5.

8.5.5 Controlling Power in RNAi Screens

For hit selection in RNAi HTS assays, we need to determine a decision rule corresponding to a critical value β^* so that we do not miss siRNAs with large effects while not including siRNAs with no or small effects. As described previously, one way to achieve this dual goal is to control both FNR and FPR. As an alternative, we may control both the power of selecting siRNAs with large effects and the power of avoiding siRNAs with no or small effects. For convenience, let us call them *power I* and *power II*, respectively. For selecting siRNAs with positive effects, power I is the probability of selecting an siRNA with $\beta \ge \beta_1$ as a hit, namely $\Pr(\text{Declare a hit given } \beta \ge \beta_1)$, and power II is the probability of not selecting an siRNA with $\beta \le \beta_2$ as a hit, namely $\Pr(\text{Not declare a hit given } \beta \le \beta_2)$. Clearly, power I $= 1 -$ FNR and power II $= 1 -$ FPR.

The power is the probability of rejecting a null hypothesis H_0 when the true value belongs to an alternative hypothesis H_a, namely $\Pr\{\text{Test rejects } H_0|\beta \in H_a\}$, which is equal to $1 -$ type II error rate. However, a power function is the plot of $\Pr\{\text{Test rejects } H_0|\beta\}$ versus β, whether β falls in H_0 or H_a, not just power versus β when β falls in H_a[149]. For convenience, let power denote $\Pr\{\text{Test rejects } H_0|\beta\}$ for any value of β, not just for β in H_a.

Considering the null hypothesis $H_0 : \beta \le \beta_2$, given a critical value β^*, $\Pr\{\text{Test rejects } H_0|\beta\} = \Pr\{\hat{\beta} > \beta^*|\beta\} = 1 - \Pr\{\hat{\beta} \le \beta^*|\beta\}$. That is,

$$\text{power} = 1 - F_{t(v,b\beta)}\left(\frac{\beta^*}{k}\right) \tag{8.43}$$

where $t(v, b\beta)$, v, b, k and $F_{t(v,b\beta)}(\cdot)$ are as in Formula 8.28. Traditionally, to get a power function, we have to fix a critical value β^*. One way to specify the critical value is to control the corresponding type I error level (which is FPL with respect to β_2) to be α, which leads to $\beta^* = kQ_{t(v,b\beta_2)}(1 - \alpha)$ where $Q_{t(v,b\beta)}(\alpha)$ is the α quantile of $t(v, b\beta)$; consequently,

$$\text{power} = 1 - F_{t(v,b\beta)}\left(Q_{t(v,b\beta_2)}(1 - \alpha)\right) \tag{8.44}$$

Clearly, the power function depends not only on β, but also on β_2. If $\beta = \beta_1$, then power $= 1 - FNL$. For $\beta_2 = 0$, power $= 1 - F_{t(v,b\beta)}\left(Q_{t(v)}(1 - \alpha)\right)$, which is a function of α and β.

Similarly, considering the null hypothesis $H_0 : \beta \geq \beta_1$, we have $\Pr\{\text{Test rejects } H_0 | \beta\} = \Pr\{\hat{\beta} < \beta^* | \beta\}$. That is

$$\text{power} = F_{t(v,b\beta_1)}\left(\frac{\beta^*}{k}\right) \tag{8.45}$$

If the corresponding type I error rate (i.e., FNL with respect to β_1) is controlled to be α, then $\beta^* = k\,Q_{t(v,b\beta_1)}(\alpha)$; consequently,

$$\text{power} = F_{t(v,b\beta)}\left(Q_{t(v,b\beta_1)}(\alpha)\right) \tag{8.46}$$

If $\beta = \beta_2$, then power $= 1 - FPL$. For $\beta_1 = 3$,

$$\text{power} = F_{t(v,b\beta)}\left(Q_{t(v,3b)}(\alpha)\right) \tag{8.47}$$

Traditional power function is the plot of power versus possible true value of β. However, our interest in hit selection is to determine a critical value or cutoff for selecting effective siRNAs. To search a critical value for selecting effective siRNAs, the best strategy is first to plot power versus critical value β^*, not β, and then to find the value of β^* that achieves a desired power. Clearly, traditional power function works neither effectively nor directly in searching β^* for selecting effective siRNAs.

If we fix β, we can get the plot of power versus β^*. For example, to select siR-NAs with positive effects, the minimal power for selecting siRNAs with $\beta \geq \beta_1$, corresponding to a critical value β^* (i.e., power I), is

$$\text{power I} = 1 - F_{t(v,b\beta_1)}\left(\frac{\beta^*}{k}\right) \tag{8.48}$$

for a fixed β_1. Similarly, the minimal power for selecting siRNAs with $\beta \leq \beta_2$ corresponding to a critical value β^* (i.e., power II) is

$$\text{power II} = F_{t(v,b\beta_2)}\left(\frac{\beta^*}{k}\right) \tag{8.49}$$

for a fixed β_2. The plots of powers versus β^* for selecting up-regulated hits with respective to $\beta_1 = 1.28, 1.645, 2, 3, 5$ and $\beta_2 = 0, 0.25$ are displayed for a screen with triplicates per siRNA in Figure 8.6A and for a screen with six replicates per siRNA in Figure 8.6B. Based on Figure 8.6, a critical value β^* between 0.5 and 1.28 may lead to reasonably high powers (i.e., power I) for selecting siRNAs with large effects (black curves) and reasonably high powers (i.e., power II) for avoiding siRNAs with small effects.

A: Powers in screens with triplicates per siRNA

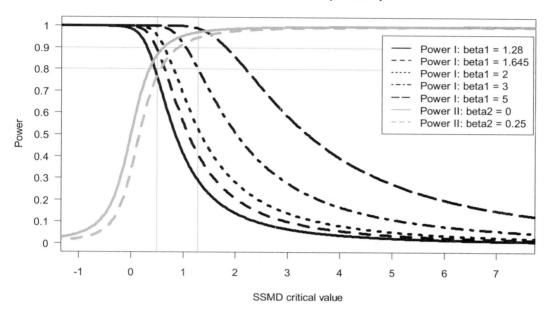

B: Powers in screens with six replicates per siRNA

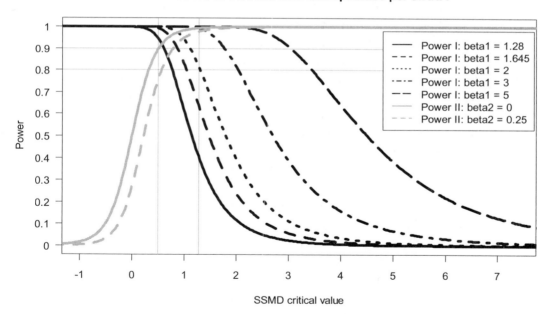

Figure 8.6   Powers for selecting siRNAs with large effects (A1, B1) and powers for avoiding siRNAs with small effects (A2, B2) in siRNAs screens.

Table 8.1. Traditional false positives and false negatives in m simultaneous tests for the construction of FDR in the up-regulated direction

	Declared Non-Significant	Declared Significant	Total
Non-interesting	U	V	m_0
$(\beta < \beta_0 \text{ or } \mu < \mu_0)$	(No. of true negatives)	(No. of false positives)	
Interesting	T	S	$m - m_0$
$(\beta \geq \beta_0 \text{ or } \mu \geq \mu_0)$	(No. of false negatives)	(No. of true positives)	
Total	$m - R$	R	m

8.6 FDR and FNDR in RNAi Screens

8.6.1 Basic Concepts of FDR

All the methods described above control FPR and FNR on the basis of a single test. Given that a large number of siRNAs are tested in a genome-scale RNAi screen, the FPR will be inflated. Hence one issue in these methods is the adjustment of error rates in multiple hypothesis testing. The simplest adjustment may be the Bonferroni correction [15], which is the FPR for a single test divided by the total number of tests conducted. Currently, the most effective method for adjusting for the multiplicity issue is the use of FDR.

Traditionally, the FDR is defined as in Table 8.1. Consider the problem of simultaneous testing m null hypotheses, of which m_0 are noninteresting (i.e., $\beta < \beta_0$ if using SSMD or $\mu < \mu_0$ if using mean difference) when the interest is in up-regulation. False positives are the siRNAs with $\beta < \beta_0$ (or $\mu < \mu_0$) among the significants and false negatives are those with $\beta \geq \beta_0$ (or $\mu \geq \mu_0$) among the non-significants. Based on Table 8.1, FPR is the expectation of the total number of false positives V divided by the total number of tests m, namely $E(\frac{V}{m})$. By contrast, FDR is the expectation of V divided by the total number of significant tests (i.e., discoveries) R, namely $E(\frac{V}{R})$. FNDR is the expectation of T divided by the total number of declared non-hits $m - R$, namely $E(\frac{T}{m-R})$.

For hit selection in a genome-scale RNAi screen, we have two goals: (i) we do not want the siRNAs with strong effects to be treated as non-hits; (ii) we do not want the siRNAs with extremely weak effects or no effects to be selected as hits. Meanwhile, we can tolerate a hit list that contains some siRNAs with weak or moderate effects. To take into account this situation in genome-scale RNAi screens, the false positives and false negatives should be defined as in Table 8.2 instead of as in Table 8.1 [169]. That is, the false positives are the siRNAs with $\beta \leq \beta_2$ (or $\mu \leq \mu_2$) among the selected hits, and the false negative are those with $\beta \geq \beta_1$ (or $\mu \geq \mu_1$) among the declared non-hits, where $\beta_2 < \beta_1$ (or $\mu_2 < \mu_1$). For example, in some RNAi screens, the false positives that we want to control are the siRNAs with SSMD ≤ 0.25 (not SSMD ≤ 3) among the selected hits; the false negatives are the siRNAs with SSMD ≥ 3 (not

Table 8.2. False positives and false negatives in m simultaneous tests for constructing FDR in the up-regulated direction in a genome-scale RNAi screen

	Declared as Non-Hits	Declared as Hits	Total
Non-interesting	U	V	m_0
($\beta \leq \beta_2$ or $\mu < \mu_2$)	(No. of true negatives)	(No. of false positives)	
Tolerable	$W1$	$W2$	m_1
($\beta_2 < \beta < \beta_1$ or $\mu_2 < \mu < \mu_1$)			
Interesting	T	S	$m - m_0 - m_1$
($\beta \geq \beta_1$ or $\mu \geq \mu_1$)	(No. of false negatives)	(No. of true positives)	
Total	$m - R$	R	m

SSMD ≥ 0.25) among the declared non-hits. We can tolerate siRNAs with SSMD between 0.25 and 3 in the selected hit list [169]. The U, V, T, and S in Table 8.2 have different meanings from those in Table 8.1. They are labeled in such a way that the definitions of FDR and FNDR based on Table 8.2 have the same format as those based on Table 8.1. That is, we still have $E(\frac{V}{R})$ for FDR and $E(\frac{T}{m-R})$ for FNDR.

There are other concepts of FDR. One of them is positive FDR, pFDR $= E(\frac{V}{R} | R > 0)$, defined as the expectation of $\frac{V}{R}$ conditional on at least one rejection [144]. Another is conditional FDR, cFDR $= E(\frac{V}{R} | R = r)/r$, defined as the expected proportion of false positives conditional on the event that $R = r$ rejected that have been observed, which answers the question "What proportion of false positives may one expect in the top list of r siRNAs?" Another method that is used less frequently is the marginal FDR, mFDR $= E(V)/E(R)$, defined as the ratio of the expected number of false positives to the expected number of rejections. Tsai, Hsueh, and Chen [152] prove that pFDR, cFDR, and mFDR are all equivalent with each other under independence and identical distribution in a Bayesian setting.

8.6.2 q-Value

The well-known q-value is a term defined similarly to p-value. The q-value is defined in terms of FDR, whereas p-value is defined in terms of FPR. Considering hit selection in the up-regulated direction, the p-value (with respect to β_2) of an siRNA with an observed value β_{obs} is p-value$(\beta_{obs}) = \max\{\text{FPR w.r.t. } \beta_2\} = \max\{\Pr(\hat{\beta} \geq \beta_{obs} | \beta \leq \beta_2)\}$. Similarly, the q-value is defined as q-value$(\beta_{obs}) = \max\{\text{FDR w.r.t. } \beta_2\}$. In terms of p_i, the q-value is q-value$(p_i) = \max_{\gamma \leq p_i}\{\text{FDR}(\gamma) \text{ w.r.t. } \beta_2\}$. When the FDR is non-increasing, as it should be, then q-value$(p_i) = \text{FDR}(p_i)$. For an individual siRNA, the q-value (with respect to β_2) of a particular siRNA with an observed value β_{obs} is the maximum FDR if we use the following selection criterion: any siRNA is selected as a hit if it has the estimated SSMD value no less than β_{obs} and as a non-hit otherwise.

8.6.3 Estimation of *q*-Value and FDR

There are an impressive number of algorithms for estimating *q*-value and/or controlling FDR in the literature [3;22;35;42;57;90;93;103;120;124;144;146]. One popular algorithm is the Benjamini–Hochberg (BH) procedure [11]. Suppose the m simultaneous tests have test statistics such as t-statistics $t_1, \ldots, t_m$ or z-scores $z_1, \ldots, z_m$ and corresponding p-values $p_1, \ldots, p_m$ respectively. The BH procedure consists of three steps: (i) order the p-values so that $p_{(1)} \leq \cdots \leq p_{(m)}$; (ii) set the desired FDR level to be q and calculate k based on $k = \max(i : p_{(i)} \leq q\frac{i}{m})$; and (iii) reject all hypotheses corresponding to $p_{(1)}, \ldots, p_{(k)}$. As proved by B, the above procedure controls FDR $\leq q$ when the test statistics are independent. The BH procedure results in a simple correction of p-values as follows, $p_i^{BH} = p_i \frac{m}{\mathrm{order}(p_i)}, i = 1, \ldots, m$, where $\mathrm{order}(p_i)$ is the rank of p_i among all the m p-values in increasing order. As compared with the Bonferroni correction, $p_i \leq p_i^{BH} \leq p_i^{Bf}$.

Mixture models are commonly used for estimating FDR. For the observed p-values, a two-component mixture model, in terms of distribution functions, is $F(p) = \eta_0 F_0(p) + (1 - \eta_0) F_A(p)$, where F_0 and F_A are the distribution functions of p-value under the null and alternative hypotheses, respectively. Under the null hypothesis, p-values are distributed with the uniform distribution $U(0, 1)$; consequently $F_0(p) = p$. Therefore, the preceding two formulas become $F(p) = \eta_0 p + (1 - \eta_0) F_A(p)$. If one declares a result significant if p-value $\leq p$ and non-significant otherwise, then the FDR is $\frac{\Pr(\text{null hypothesis and p.value } \leq p)}{\Pr(\text{p.value} \leq p)} = \frac{\eta_0 F_0(p)}{F(p)} = \frac{\eta_0 p}{F(p)}$. Based on Bayes rule, $\Pr(\text{null hypothesis}|\text{p.value } \leq p) = \frac{\Pr(\text{null hypothesis and p.value } \leq p)}{\Pr(\text{p.value} \leq p)}$. Therefore, FDR is actually the posterior distribution for the null hypothesis, which leads to the Bayesian definition of FDR: $\mathrm{FDR} = \Pr(\text{null hypothesis}|\text{p.value } \leq p) = \frac{\eta_0 p}{F(p)}$ [145]. The Bayesian FDR can be used to interpret BH-corrected p-value as follows. For observed p-value p_i, $\mathrm{FDR}(p_i) = \Pr(\text{null hypothesis}|\text{p.value } \leq p_i) = \frac{\eta_0 p_i}{F(p_i)}$. Using the empirical distribution function $\hat{F}(p_i) = \frac{\mathrm{order}(p_i)}{m}$ and considering $\hat{\eta}_0 \leq 1$, one gets $\hat{\mathrm{FDR}}(p_i) = \eta_0 p_i \frac{m}{\mathrm{order}(p_i)} \leq p_i \frac{m}{\mathrm{order}(p_i)} = p_i^{BH}$. Thus the BH-corrected p-value is a conservative estimator of FDR.

All the above FDRs are based on distribution function. This type of FDRs is called *tail area–based FDR*. By contrast, FDR can also be defined on the basis of density function as $\mathrm{FDR} = \Pr(\text{null hypothesis}|\text{p.value } = p)$. This type of FDR is called *local FDR*. In terms of density functions, a two-component mixture model based on p-values is $f(p) = \eta_0 f_0(p) + (1 - \eta_0) f_A(p)$, where f_0 and f_A are the density functions of p-value under the null and alternative hypotheses, respectively. Under the null hypothesis, p-values are distributed with the uniform distribution $U(0, 1)$; consequently, $f_0 = 1$ and $f(p) = \eta_0 + (1 - \eta_0) f_A(p)$. Thus $\mathrm{FDR} = \frac{\eta_0 f_0(p)}{f(p)} = \frac{\eta_0}{f(p)}$.

Therefore, there are two main types of FDR, the tail area–based FDR and local FDR. Following Efron's naming convention, *fdr* denotes the local FDR, *Fdr* denotes tail area–based FDR, and *FDR* is a generic term encompassing both variants. Using this convention, we have $\mathrm{Fdr} = \frac{\eta_0 F_0(p)}{F(p)} = \frac{\eta_0 p}{F(p)}$ and $\mathrm{fdr} = \frac{\eta_0 f_0(p)}{f(p)} = \frac{\eta_0}{f(p)}$. To estimate either Fdr or fdr, the key is to estimate the density function $f(p)$ (or the distribution function $F(p)$) and the proportion η_0 of true null hypotheses. There are many approaches to estimate $F(p)$ and η_0 [3;21;22;35;42;57;90;93;103;124;144;146;147].

8.6.4 FNDR and q^*-Value

As parallel to FDR, FNDR) is defined as the ratio of the number of false negatives to the number of all negatives, namely the expectation of T divided by the total number of non-significant tests (i.e., non-discoveries) $m - R$, i.e., $\mathrm{E}(\frac{T}{m-R})$ based on Table 8.2 or Table 8.1 [52]. Similarly, as FDR corresponds to FPR, FNDR corresponds to FNR. In traditional hypothesis testing, researchers are more concerned with FDR than with FNDR. However, in SSMD-based tests, region I1 may be treated as equally important as region I2 in Figure 8.4. In such a case, FNDR can be equally as important as FDR.

In the context of FDR, corresponding to p-value, we have q-value. In parallel, in the context of FNDR, corresponding to p^*-value, we have q^*-value. Considering hit selection in the up-regulated direction, the p^*-value (with respect to β_1) of an siRNA with an observed value β_{obs} is p^*-value$(\beta_{\mathrm{obs}}) = \max\{\mathrm{FNR\ w.r.t.}\ \beta_1\} = \max\{\Pr(\hat{\beta} < \beta_{\mathrm{obs}}|\beta \geq \beta_1)\}$. Similarly, the q^*-value is defined as q^*-value$(\beta_{\mathrm{obs}}) = \max\{\mathrm{FNDR\ w.r.t.}\ \beta_1\}$. The q^*-value has the following meaning for an individual siRNA: the q^*-value (with respect to β_1) of a particular siRNA with an observed value β_{obs} is the maximum FNDR if we use the following selection criterion: any siRNA is selected as a hit if it has the estimated SSMD value no less than β_{obs} and as a non-hit otherwise.

Using the methods described in Section 8.6.3, we can easily calculate the q-value corresponding to the p-value for testing $\mathrm{H}_0 : \beta \leq \beta_2$ (or $\mathrm{H}_0 : \mu \leq \mu_2$). To calculate q^*-value, we can treat p^*-value as the p-value for testing $\mathrm{H}_0 : \beta \geq \beta_1$ (or $\mathrm{H}_0 : \mu \geq \mu_1$) and calculate the corresponding q-value; the resulting q-value equals the q^*-value with respect to $\beta \geq \beta_1$ (or $\mu \geq \mu_1$).

8.7 Analytic Methods Adjusting for Off-Target Effects

8.7.1 Introduction to Off-Target Effects

Off-target effects occur when an siRNA is processed by the RNA-induced silencing complex (RISC) and down-regulates unintended targets. Although other non-specific effects, such as lipid-mediated response and interferon response, can be eliminated by adopting stringent siRNA design filters and optimizing lipid concentrations and compositions, off-target effects pose a bigger challenge by presenting the research community with a surprisingly complex problem [78;80;130;133;137]. During siRNA design stage, siRNA pooling, siRNA modification [79], and 3'UTR seed match [12;78;94] can be used to reduce off-target effects during siRNA design.

In the experimental stage, the main approach to adjusting for off-target effects is the use of multiple siRNAs targeting the same gene. There are many approaches and designs for reducing the impact of off-target effects. The most common and accessible approach is the conduction of the so-called deconvolution screen, in which multiple siRNAs are tested separately with different sequences against a target gene to increase the level of confidence in positive hits [14;40]. The major reason for examining the collective activity of multiple siRNAs is that the off-target effects of these siRNAs are very likely to have different directions and thus may be canceled out in their collective activity, whereas the on-target effects of these siRNAs should be in the same direction and thus may be accumulated (or at least will not be canceled out with each other) in their collective activity. Accordingly, analytic methods have to incorporate the information of multiple siRNAs targeting the same gene. Here we are modeling the collective activity of multiple siRNAs with different sequences against a target gene.

8.7.2 Model for Collective Activity of Multiple siRNAs

Suppose we are interested in the collective activities of m siRNAs that are measured separately. The activity of an individual siRNA on a measured response is usually represented by the difference of measured values between this siRNA and a negative reference group. Let d_{ij} denote the difference in the jth replicate of the ith siRNA and its corresponding negative reference. Assuming the ith siRNA has a mean value of μ_i, we can construct the following model for d_{ij}:

$$d_{ij} = \mu_i + e_{ij} \tag{8.49}$$

where $i = 1, \ldots, m$ and $j = 1, \ldots, n_i$; e_{ij}'s are independently distributed with $N(0, \sigma_{i.e}^2)$; and μ_i is mean of the ith siRNA, which includes the on-target and off-target effect that the siRNA has (i.e., $\mu_i = \mu_{i.\text{on-target}} + \mu_{i.\text{off-target}}$).

The collective activity of m siRNAs is represented by the weighted average activity of m values each drawn from one of the m siRNAs. Thus we can investigate it as follows. For the ith siRNA among the m siRNAs, let random variable D_i represent the difference of measured values between the siRNA and a negative reference. D_i has a density function f_i, mean μ_i, and variance σ_i^2. The collective effect of the m siRNAs is represented by the difference $D_{\text{collective}}$ of measured values between an siRNA and a negative reference in a group that is formed by pooling all the m siRNAs with weight w_i's, that is, $f_{\text{collective}} = \sum_{i=1}^{m} (w_i f_i)$, where $f_{\text{collective}}$ is the density function of $D_{\text{collective}}$. The weight w_i's have a constraint of $\sum_{i=1}^{m} w_i = 1$. In many cases, the m siRNAs have equal weights, that is, $w_i = \frac{1}{m}$. Let $\mu_{\text{collective}}$ and $\sigma_{\text{collective}}^2$ be the mean and variance of $D_{\text{collective}}$, respectively, and $\mu_\bullet = \sum_{i=1}^{m} w_i \mu_i$. Then $\mu_{\text{collective}} = \int x \cdot f_{\text{collective}} dx = \int x \cdot \sum_{i=1}^{m} (w_i f_i) dx = \sum_{i=1}^{m} w_i \int x f_i dx = \sum_{i=1}^{m} w_i \mu_i$. Therefore,

$$\mu_{\text{collective}} = \mu_\bullet = \sum_{i=1}^{m} w_i \mu_i \tag{8.50}$$

The mean of the ith siRNA, μ_i, includes its on-target effect $\mu_{i.\text{on-target}}$ and off-target effect $\mu_{i.\text{off-target}}$. It is well-known that one siRNA has a consistently strong phenotypic effect on a gene and another siRNA has a consistently weak phenotypic effect on the same gene, regardless of the RNAi libraries used. Therefore, siRNAs targeting the same gene may have different specific on-target effects beyond the on-target effect shared by all the m siRNAs. Consequently, we can partition the on-target effect of the ith siRNA $\mu_{i.\text{on-target}}$ into two parts: the shared on-target effect $\mu_{i.\text{shared on-target}}$ and specific on-target effect $\mu_{i.\text{specific on-target}}$ (i.e., $\mu_{i.\text{on-target}} = \mu_{i.\text{shared on-target}} + \mu_{i.\text{specific on-target}}$). On the basis of this partition, we have $\mu_{\text{collective}} = \sum_{i=1}^{m} w_i (\mu_{i.\text{on-target}} + \mu_{i.\text{off-target}})$ $= \sum_{i=1}^{m} w_i (\mu_{\text{shared-on-target}} + \mu_{i.\text{specific-on-target}} + \mu_{i.\text{off-target}}) = \mu_{\text{shared-on-target}} + \bar{\mu}_{\text{specific-on-target}} + \bar{\mu}_{\text{off-target}}$. Thus $\mu_{\text{collective}}$ is the shared on-target effect plus the sum of specific on-target and off-target effects weightily averaged over the m siRNAs. When m is large, the off-target effect should be canceled out, as should be the specific on-target effect. That is, $\bar{\mu}_{\text{specific-on-target}} + \bar{\mu}_{\text{off-target}} \approx 0$; subsequently, $\mu_{\text{collective}} \approx \mu_{\text{shared-on-target}}$, which indicates that the mean of collective activity of a large number of siRNAs targeting the same gene represents the shared on-target effects of these siRNAs on the gene.

For the variance of $D_{\text{collective}}$, we have $\sigma^2_{\text{collective}} = \int (x - \mu_{\text{collective}})^2 f_{\text{collective}} dx = \int (x - \mu_{\text{collective}})^2 \cdot \sum_{i=1}^{m} w_i f_i dx = \sum_{i=1}^{m} w_i \int (x - \mu_i + \mu_i - \mu_{\text{collective}})^2 f_i dx = \sum_{i=1}^{m} (w_i \int (x - \mu_i)^2 f_i dx) + \sum_{i=1}^{m} w_i (\mu_i - \mu_{\text{collective}})^2 = \sum_{i=1}^{m} w_i \sigma_i^2 + \sum_{i=1}^{m} w_i (\mu_i - \mu_{\text{collective}})^2$. Clearly, the variance of $D_{\text{collective}}$ consists of two parts: one is $\sum_{i=1}^{m} w_i \sigma_i^2$, contributed by within-siRNA variation, and the other is $\sum_{i=1}^{m} w_i (\mu_i - \mu_{\text{collective}})^2$, contributed by between-siRNA variation. The within-siRNA variation comes from the variation of technical replicates of the same siRNAs; thus it represents the technical or measurement variation. Meanwhile, if we treat different siRNAs targeting the same gene as a biological replicate, the between-siRNA variation represents the biological variation. For convenience, let's use $\sigma_{\text{technical}}$ to denote the within-siRNA variation and use $\sigma_{\text{biological}}$ to denote the between-siRNA variation. Then the SSMD for the collective activity based on biological variation is

$$\beta_{\text{biological}} = \frac{\mu_{\text{collective}}}{\sigma_{\text{biological}}} = \frac{\sum_{i=1}^{m} w_i \mu_i}{\sqrt{\sum_{i=1}^{m} w_i (\mu_i - \mu_\bullet)^2}}. \tag{8.51}$$

The SSMD for the collective activity based on technical variation is

$$\beta_{\text{technical}} = \frac{\mu_{\text{collective}}}{\sigma_{\text{technical}}} = \frac{\sum_{i=1}^{m} w_i \mu_i}{\sqrt{\sum_{i=1}^{m} w_i \sigma_i^2}}. \tag{8.52}$$

The SSMD for the collective activity based on both biological and technical variation is

$$\beta_{\text{both}} = \frac{\mu_{\text{collective}}}{\sigma_{\text{both}}} = \frac{\sum\limits_{i=1}^{m} w_i \mu_i}{\sqrt{\sum\limits_{i=1}^{m} w_i \sigma_i^2 + \sum\limits_{i=1}^{m} w_i (\mu_i - \mu_\bullet)^2}}. \tag{8.53}$$

When the independent D_i's have equal variance σ^2, then the SSMD for the collective activity based on technical variation and on both biological and technical variation are, respectively,

$$\beta_{\text{technical}} = \frac{\mu_{\text{collective}}}{\sigma_{\text{technical}}} = \frac{\sum\limits_{i=1}^{m} w_i \mu_i}{\sigma} \tag{8.54}$$

$$\beta_{\text{both}} = \frac{\mu_{\text{collective}}}{\sigma_{\text{both}}} = \frac{\sum\limits_{i=1}^{m} w_i \mu_i}{\sqrt{\sigma^2 + \sum\limits_{i=1}^{m} w_i (\mu_i - \mu_\bullet)^2}} \tag{8.55}$$

Biologists are usually more interested in biological replicates than in technical replicates. Therefore, the SSMD based on biological variation in Formula 8.51 is usually favorable.

8.7.3 Estimation of Mean and SSMD for the Collective Activity of Multiple siRNAs

Let $D_i = (d_{i1}, d_{i2}, \ldots, d_{in_i})$, $i = 1, \ldots, m$ be a random sample of the difference in the ith siRNA. Let n_i, $\bar{d}_i$, and s_i^2 be the sample size, sample mean, and sample variance of D_i, respectively. Let $\bar{d}_\bullet = \sum_{i=1}^{m} w_i \bar{d}_i$, $n = \sum_{i=1}^{m} n_i$ and MSE $= \frac{1}{n-m} \sum_{i=1}^{m} (n_i - 1) s_i^2$. Based on formulas 8.50, 8.51, 8.52, 8.53, 8.54, and 8.55, the mean and SSMD of the collective activity $D_{\text{collective}}$ for the m siRNAs targeting the same gene can be estimated using the following MM) estimates:

$$\mu_{\text{collective}} = \bar{d}_\bullet \tag{8.56}$$

$$\hat{\beta}_{\text{biological}} = \frac{\bar{d}_\bullet}{\sqrt{\sum\limits_{i=1}^{m} w_i (\bar{d}_i - \bar{d}_\bullet)^2}} \tag{8.57}$$

$$\hat{\beta}_{\text{technical}} = \frac{\bar{d}_\bullet}{\sqrt{\sum\limits_{i=1}^{m} w_i s_i^2}} \tag{8.58}$$

$$\hat{\beta}_{\text{both}} = \frac{\bar{d}_{\bullet}}{\sqrt{\sum_{i=1}^{m} w_i s_i^2 + \sum_{i=1}^{m} w_i (\bar{d}_i - \bar{d}_{\bullet})^2}} \tag{8.59}$$

When the independent D_i's have equal variance σ^2,

$$\hat{\beta}_{\text{technical}} = \frac{\bar{d}_{\bullet}}{\sqrt{\text{MSE}}} \tag{8.60}$$

$$\hat{\beta}_{\text{both}} = \frac{\bar{d}_{\bullet}}{\sqrt{\text{MSE} + \sum_{i=1}^{m} w_i (\bar{d}_i - \bar{d}_{\bullet})^2}} \tag{8.61}$$

8.7.4 Individual Activity of an Individual siRNA

The individual activity of an siRNA (siRNA i) can be assessed using the random variable D_i representing the difference between siRNA i and a negative reference. It is trivial to make estimation and inference for the mean of D_i. On the basis of formulas 8.11, 8.12, 8.13, and 8.14, the estimation and inference of SSMD for D_i can be obtained using the following formulas:

$$\hat{\beta}_{i.\text{UMVUE}} = \frac{\Gamma\left(\dfrac{n_i - 1}{2}\right)}{\Gamma\left(\dfrac{n_i - 2}{2}\right)} \sqrt{\frac{2}{n_i - 1}} \frac{\bar{d}_i}{s_i} \tag{8.62}$$

$$\hat{\beta}_{i.\text{MLE}} = \sqrt{\frac{n_i}{n_i - 1}} \frac{\bar{d}_i}{s_i}, \tag{8.63}$$

$$\hat{\beta}_{i.\text{MM}} = \frac{\bar{d}_i}{s_i}, \tag{8.64}$$

$$T_i = \sqrt{n_i} \frac{\bar{d}_i}{s_i} \sim \text{noncentral } t(n_i - 1, \sqrt{n_i}\beta_i). \tag{8.65}$$

The $1 - \alpha$ confidence interval of μ_i is $\bar{d}_i \pm t_{n_i-1, \frac{\alpha}{2}} \frac{s_i}{\sqrt{n_i}}$.

8.7.5 Specific Activity of an Individual siRNA

The specific effect of an individual siRNA beyond the shared on-target effect may be caused by either off-target or specific on-target effects. This specific effect is represented by the magnitude of difference between the individual siRNA and all siRNAs targeting the same gene. This magnitude of difference can be addressed using the SMCV and mean of the contrast for the main effect of the siRNA when we treat each siRNA targeting the same gene as a factor level in one-way ANOVA. The specific effect of the ith siRNA among the m siRNAs targeting the same gene can then be assessed using SMCV of the contrast variable:

$$V_i = D_i - D_{\bullet} = D_i - \sum_{k=1}^{m} w_k D_k = (1 - w_i)D_i + \sum_{k \neq i}^{m} (-w_k D_k) = \sum_{k=1}^{m} c_k D_k$$

where $c_k = \begin{cases} 1 - w_k, & \text{when } k = i \\ -w_k, & \text{when } k \neq i \end{cases}$. Applying Theorem 2 in Section 7.4 of Chapter 7, we can make estimation and inference of mean and SMCV of the contrast variable V_i for addressing the specific effect of siRNA i as follows.

The mean of V_i is

$$\begin{aligned}
\tau_i &= \mu_i - \sum_{k=1}^{m} w_k \mu_k \\
&= \mu_{\text{shared-on-target}} + \mu_{i.\text{specific-on-target}} + \mu_{i.\text{off-target}} \\
&\quad - \sum_{k=1}^{m} w_k \left(\mu_{\text{shared-on-target}} + \mu_{k.\text{specific-on-target}} + \mu_{k.\text{off-target}} \right) \\
&= \mu_{i.\text{specific-on-target}} + \mu_{i.\text{off-target}} - \sum_{k=1}^{m} w_k \left(\mu_{k.\text{specific-on-target}} + \mu_{k.\text{off-target}} \right)
\end{aligned}$$

Therefore, τ_i is a combination of specific on-target and off-target effects of the siRNA away from the sum of specific on-target and off-target effects weightily averaged over the m investigated siRNAs. If the sum of specific on-target and off-target effects weightily averaged over the m investigated siRNAs is zero, that is, $\sum_{k=1}^{m} w_k \left(\mu_{k.\text{specific-on-target}} + \mu_{k.\text{off-target}} \right) = 0$, which approximately holds especially when the m is large, then $\tau_i = \mu_{i.\text{specific-on-target}} + \mu_{i.\text{off-target}}$. Thus τ_i roughly represents the sum of specific on-target and off-target effect of the ith siRNA.

Whether the m siRNAs have equal or unequal variances, the estimate of τ_i is estimated to be

$$\hat{\tau}_i = \sum_{k=1}^{m} c_k \bar{d}_k. \tag{8.66}$$

In a situation in which the m siRNAs have unequal variance, the MLE of SMCV for V_i (from Formula T2.1 in Chapter 7) is

$$\hat{\lambda}_{i.\text{MLE}} = \frac{\sum_{k=1}^{m} c_k \bar{d}_k}{\sqrt{\sum_{k=1}^{m} \frac{n_k - 1}{n_k} c_k^2 s_k^2}}; \tag{8.67}$$

the MM of SMCV for V_i (from Formula T2.2 in Chapter 7) is

$$\hat{\lambda}_{i.\text{MM}} = \frac{\sum\limits_{k=1}^{m} c_k \bar{d}_k}{\sqrt{\sum\limits_{k=1}^{m} c_k^2 s_k^2}}. \tag{8.68}$$

In a situation with unequal variance but equal sample size r for the m siRNAs, an approximate unbiased estimate of SMCV for V_i (from Formula T2.5 in Chapter 7) is

$$\hat{\lambda}_{i.\text{AUE}} = \sqrt{\frac{2}{\nu}} \frac{\Gamma\left(\frac{\nu}{2}\right)}{\Gamma\left(\frac{\nu-1}{2}\right)} \frac{\sum\limits_{k=1}^{m} c_k \bar{d}_k}{\sqrt{\sum\limits_{k=1}^{m} c_k^2 s_k^2}}, \text{ where } \nu = (r-1)\frac{\left(\sum\limits_{k=1}^{m} c_k^2 s_k^2\right)^2}{\sum\limits_{k=1}^{m} c_k^4 s_k^4}. \tag{8.69}$$

In a situation in which the m siRNAs have equal variance, the $1 - \alpha$ confidence interval of the mean of V_i is

$$\sum_{k=1}^{m} c_k \bar{d}_k \pm t_{1-\alpha/2, N-m} \times \sqrt{\text{MSE} \cdot \sum_{k=1}^{m} c_k^2 / n_k^2}, \tag{8.70}$$

where $N = \sum_{k=1}^{m} n_k$ and $\text{MSE} = \frac{1}{N-m} \sum_{k=1}^{m} (n_k - 1)s_k^2$.

The UMVUE, MLE, and MM estimates of λ_i are, respectively (from Formulas T2.7–T2.9 in Chapter 7),

$$\hat{\lambda}_{i.\text{UMVUE}} = \frac{\sqrt{K}}{\sqrt{N-m}} \frac{\sum\limits_{k=1}^{m} c_k \bar{d}_k}{\sqrt{\text{MSE} \cdot \sum\limits_{k=1}^{m} c_k^2}}, \text{ where } K = 2 \cdot \left(\frac{\Gamma\left(\frac{N-m}{2}\right)}{\Gamma\left(\frac{N-m-1}{2}\right)}\right)^2 \tag{8.71}$$

$$\hat{\lambda}_{i.\text{MLE}} = \sqrt{\frac{N}{N-m}} \frac{\sum\limits_{k=1}^{m} c_k \bar{d}_k}{\sqrt{\text{MSE} \cdot \sum\limits_{k=1}^{m} c_k^2}} \tag{8.72}$$

$$\hat{\lambda}_{i.\text{MM}} = \frac{\sum\limits_{k=1}^{m} c_k \bar{d}_k}{\sqrt{\text{MSE} \cdot \sum\limits_{k=1}^{m} c_k^2}} \tag{8.73}$$

From Formula T2.9 in Chapter 7, we have $T = (\sum_{i=1}^{m} c_k \bar{d}_k)/$ $\left(\sqrt{\text{MSE}} \cdot \sum_{k=1}^{m} c_k^2 / n_k\right) \sim$ noncentral $t(\nu, b\lambda_i)$, where $\nu = N - m$ and

$b = \sqrt{(\sum_{k=1}^{m} c_k^2)/(\sum_{k=1}^{m} c_k^2/n_k)}$. Let $F_{t(\nu, b\lambda_i)}(\cdot)$ be the cumulative distribution function of noncentral $t(\nu, b\lambda_i)$ and T_{obs} be the observed value of T. Then we can find λ_L and λ_U such that $F_{t(\nu, b\lambda_L)}(T_{\text{obs}}) = 1 - \frac{\alpha}{2}$ and $F_{t(\nu, b\lambda_U)}(T_{\text{obs}}) = \frac{\alpha}{2}$; subsequently, (λ_L, λ_U) is a $1 - \alpha$ confidence interval of SMCV λ_i.

8.8 Discussion and Conclusions

In most genome-scale RNAi screens, the ultimate goal is to select siRNAs with a desired size of inhibition or activation effect. Traditionally, hit selection is based on the test of mean difference. However, mean difference can neither take into account data variability nor accommodate different measurement units. Consequently, the value of mean difference is not comparable across experiments, and hence no cutoff of mean difference can be applicable to various experiments. An alternative that can avoid these issues of mean difference is the so-called effect size [84]. One effect size that has been developed for HTS experiments is SSMD [161;162;167]. SSMD is the ratio of mean to standard deviation of the difference between an siRNA and a negative reference group. SSMD has also been shown to be better than other commonly used effect sizes [171]. In this chapter, I first elaborate how to derive SSMD, d^+-probability, and their estimation based on the concepts and theorems described in Chapter 7, then compare SSMD with classical t-statistic and standardized mean difference, including Cohen's d.

A clear advantage of SSMD over mean difference is that the population value of SSMD is comparable across experiments; thus we can use the same cutoff for the population value of SSMD to measure the size of siRNA effects [161;167;175]. Derived from Table 7.2 of Chapter 7, a meaningful and interpretable SSMD-based criterion for classifying the size of siRNA effects is as follows: $|\text{SSMD}| \geq 5$ for extremely strong, $5 > |\text{SSMD}| \geq 3$ for very strong, $3 > |\text{SSMD}| \geq 2$ for strong, $2 > |\text{SSMD}| \geq 1.645$ for fairly strong, $1.645 > |\text{SSMD}| \geq 1.28$ for moderate, $1.28 > |\text{SSMD}| \geq 1$ for fairly moderate, $1 > |\text{SSMD}| \geq 0.75$ for fairly weak, $0.75 > |\text{SSMD}| > 0.5$ for weak, $0.5 \geq |\text{SSMD}| > 0.25$ for very weak, and $|\text{SSMD}| \leq 0.25$ for extremely weak effects [167]. This SSMD-based criterion not only provides us with a theoretical basis for using SSMD as both an effect size metric to gauge the size of siRNA effects and a quality assessment metric to assess the separation between positive and negative controls in an assay, but also allows us to set up meaningful constants for controlling false positives and false negatives.

The hit selection usually requires the control of two errors: false positives and false negatives, which is commonly achieved through FPR or p-value [13;17;161;175], FDR or q-value [11;146;176], FNR [178], and FNDR [52]. The FPR, FDR, FNR, and FNDR based on traditional definitions of false positives and false negatives are inappropriate to serve the need of controlling the proportion of siRNAs with a small effect among selected hits and controlling the proportion of siRNAs with a large effect among declared non-hits. To address this need, in this chapter I elaborate recently proposed definitions of false positives and false negatives and their corresponding methods, including p-value, q-value, p^*-value, and q^*-value based on SSMD. The SSMD-based methods apply two constants to define non-interesting, tolerable, and interesting siRNAs (shown in Table 8.2), compared with a single constant in traditional methods (shown in Table 8.1). For example, in some RNAi screens, the false positives that we want to control are the siRNAs with SSMD ≥ -0.25 among hits; the false negatives are the siRNAs with SSMD ≤ -3 among non-hits. We can tolerate siRNAs with SSMD between -0.25 and -3 in the hit list. Thus the SSMD-based methods appropriately address the scientific need for hit selection in RNAi screens.

Off-target effects may impede the analysis of RNAi screens because false positives generated by off-targets during phenotypic screens can lead to false leads and the use of resources to explore nonproductive research paths. In this chapter, I introduce an analytic method using the average fold change and collective SSMD to select hits addressing for off-target effects in a deconvolution screen in which multiple single siRNAs are measured separately against a gene. This method naturally incorporates all the information of multiple siRNAs targeting the same gene in a strong statistical basis. In addition, this method can assess not only the collective activities of multiple siRNAs against a gene, but also the strength of specific effect of each siRNA beyond its collective activity. The consideration of both collective activity and specific activity can also give a reference about which siRNAs are more likely to have large off-target effects and which siRNAs are more likely to have specific on-target effects.

All the methods in this chapter have been derived from a methodological perspective based on scientific needs in genome-scale RNAi screens. The practical usefulness of the proposed methods in real genome-scale RNAi screens are shown in Chapters 2 through 6. Although the methods presented in this book were developed for hit selection in RNAi-based high throughput screens, they should be applicable to other assays in which the end point is a difference in signal compared with a reference sample, including those for receptors, enzymes, and cellular function.

References

[1] L. Acion, J.J. Peterson, S. Temple, S. Arndt. Probabilistic index: an intuitive non-parametric approach to measuring the size of treatment effects. *Statistics in Medicine* 25 (2006) 591–602.

[2] T.W. Anderson, D.A. Darling. Asymptotic theory of certain "goodness-of-fit" criteria based on stochastic processes. *Annals of Mathematical Statistics* 23 (1952) 193–212.

[3] J. Aubert, A. Bar-Hen, J.J. Daudin, S. Robin. Determination of the differentially expressed genes in microarray experiments using local FDR. *BMC Bioinformatics* 5 (2004) 125.

[4] D. Bakan. The test of significance in psychological research. *Psychological Bulletin* 66 (1966) 423–37.

[5] P. Baldi, G.W. Hatfield. *DNA Microarrays and Gene Expression.* Cambridge, UK, Cambridge University Press, 2002.

[6] D.A. Barbie, P. Tamayo, J.S. Boehm, S.Y. Kim, S.E. Moody, I.F. Dunn, A.C. Schinzel, P. Sandy, E. Meylan, C. Scholl, S. Frohling, E.M. Chan, M.L. Sos, K. Michel, C. Mermel, S.J. Silver, B.A. Weir, J.H. Reiling, Q. Sheng, P.B. Gupta, R.C. Wadlow, H. Le, S. Hoersch, B.S. Wittner, S. Ramaswamy, D.M. Livingston, D.M. Sabatini, M. Meyerson, R.K. Thomas, E.S. Lander, J.P. Mesirov, D.E. Root, D.G. Gilliland, T. Jacks, W.C. Hahn. Systematic RNA interference reveals that oncogenic KRAS-driven cancers require TBK1. *Nature* 462 (2009) 108–22.

[7] F. Bard, L. Casano, A. Mallabiabarrena, E. Wallace, K. Saito, H. Kitayama, G. Guizzunti, Y. Hu, F. Wendler, R. DasGupta, N. Perrimon, V. Malhotra. Functional genomics reveals genes involved in protein secretion and Golgi organization. *Nature* 439 (2006) 604–7.

[8] G.A. Barkerand, S.L. Diamond. RNA interference screen to identify pathways that enhance or reduce nonviral gene transfer during lipofection. *Molecular Therapy* 16 (2008) 1602–8.

[9] T. Barrett, D.B. Troup, S.E. Wilhite, P. Ledoux, D. Rudnev, C. Evangelista, I.F. Kim, A. Soboleva, M. Tomashevsky, R. Edgar. NCBI GEO: mining tens of millions of expression profiles – database and tools update. *Nucleic Acids Research* 35 (2007) D760–5.

[10] R.A. Becker, J.M. Chambers, A.R. Wilks. *The New S Language.* Pacific Grove, CA, Wadsworth & Brooks/Cole, 1988.

[11] Y. Benjamini, Y. Hochberg. Controlling the false discovery rate – A practical and powerful approach to multiple testing. *Journal of the Royal Statistical Society Series B-Methodological* 57 (1995) 289–300.

[12] A. Birmingham, E.M. Anderson, A. Reynolds, D. Ilsley-Tyree, D. Leake, Y. Fedorov, S. Baskerville, E. Maksimova, K. Robinson, J. Karpilow, W.S. Marshall, A. Khvorova. 3′ UTR seed matches, but not overall identity, are associated with RNAi off-targets. *Nature Methods* 3 (2006) 199–204.

[13] A. Birmingham, L.M. Selfors, T. Forster, D. Wrobel, C.J. Kennedy, E. Shanks, J. Santoyo-Lopez, D.J. Dunican, A. Long, D. Kelleher, Q. Smith, R.L. Beijersbergen, P. Ghazal, C.E. Shamu. Statistical methods for analysis of high-throughput RNA interference screens. *Nature Methods* 6 (2009) 569–75.

[14] N. Blow. RNAi technologies: a screen whose time has arrived. *Nature Methods* 5 (2008) 361–6.

[15] C.E. Bonferroni. Il calcolo delle assicurazioni su gruppi di teste. In *Studi in Onore del Professore Salvatore Ortu Carboni*. Rome, Italy, 1935, pp. 13–60.

[16] M. Boutros, J. Ahringer. The art and design of genetic screens: RNA interference. *Nature Reviews Genetics* 9 (2008) 554–66.

[17] M. Boutros, L.P. Bras, W. Huber. Analysis of cell-based RNAi screens. *Genome Biology* 7 (2006) R66.

[18] M. Boutros, A.A. Kiger, S. Armknecht, K. Kerr, M. Hild, B. Koch, S.A. Haas, R. Paro, N. Perrimon. Genome-wide RNAi analysis of growth and viability in Drosophila cells. *Science* 303 (2004) 832–5.

[19] A.L. Brass, D.M. Dykxhoorn, Y. Benita, N. Yan, A. Engelman, R.J. Xavier, J. Lieberman, S.J. Elledge. Identification of host proteins required for HIV infection through a functional genomic screen. *Science* 319S (2008) 921–6.

[20] C. Brideau, B. Gunter, B. Pikounis, A. Liaw. Improved statistical methods for hit selection in high-throughput screening. *Journal of Biomolecular Screening* 8 (2003) 634–47.

[21] P. Broberg. Statistical methods for ranking differentially expressed genes. *Genome Biology* 4 (2003), R41.

[22] P. Broberg. A comparative review of estimates of the proportion unchanged genes and the false discovery rate. *BMC Bioinformatics* 6 (2005) 199.

[23] K. Brown, D. Samarsky. RNAi off-targeting: Light at the end of the tunnel. *Journal of RNAi and Gene Silencing* 2 (2006) 175–7.

[24] D. Bumcrot, M. Manoharan, V. Koteliansky, D.W.Y. Sah. RNAi therapeutics: a potential new class of pharmaceutical drugs. *Nature Chemical Biology* 2 (2006) 711–19.

[25] Y. Cao, R.M. Kumar, B.H. Penn, C.A. Berkes, C. Kooperberg, L.A. Boyer, R.A. Young, S.J. Tapscott. Global and gene-specific analyses show distinct roles for Myod and Myog at a common set of promoters. *EMBO Journal* 25 (2006) 502–11.

[26] R.P. Carver. Case against statistical significance testing. *Harvard Educational Review* 48 (1978) 378–99.

[27] P. Cherepanov, G. Maertens, P. Proost, B. Devreese, J. Van Beeumen, Y. Engelborghs, E. De Clercq, Z. Debyser. HIV-1 integrase forms stable tetramers and associates with LEDGF/p75 protein in human cells. *Journal of Biological Chemistry* 278 (2003) 372–81.

[28] N.J. Chung, X.D. Zhang, A. Kreamer, L. Locco, P.F. Kuan, S. Bartz, P.S. Linsley, M. Ferrer, B. Strulovici. Median absolute deviation to improve hit selection for genome-scale RNAi screens. *Journal of Biomolecular Screening* 13 (2008) 149–58.

[29] J.D. Church, B. Harris. The estimation of reliability from stress-strength relationships. *Technometrics* 12 (1970) 49–54.

[30] W.S. Cleveland. *The Elements of Graphing Data*. Summit, NJ, Hobart Press, 1994.

[31] W.S. Cleveland, E. Grosse, W.M. Shyu. Local regression models. In: J.M. Chambers and T.J. Hastie (Eds.), *Statistical Models in S*. Pacific Grove, CA, Wadsworth & Brooks/Cole, 1992.

[32] J. Cohen. The statistical power of abnormal-social psychological research: A review. *Journal of Abnormal and Social Psychology* 65 (1962) 145–53.

[33] J. Cohen. *Statistical Power Analysis for the Behavioral Sciences*. Erlbaum, Hillsdale, NJ, 1988.

[34] J. Cohen. The earth is round (P less than .05). *American Psychologist* 49 (1994) 997–1003.

[35] C. Dalmasso, P. Broet, T. Moreau. A simple procedure for estimating the false discovery rate. *Bioinformatics* 21 (2005) 660–8.

[36] R. Davidson, J.Y. Duclos. Statistical inference for stochastic dominance and for the measurement of poverty and inequality. *Econometrica* 68 (2000) 1435–64.

[37] F. Downton. The estimation of $Pr(Y < X)$ in the normal case. *Technometrics* 15 (1973) 551–8.

[38] S. Draghici, P. Khatri, A.L. Tarca, K. Amin, A. Done, C. Voichita, C. Georgescu, R. Romero. A systems biology approach for pathway level analysis. *Genome Research* 17 (2007) 1537–45.

[39] B.J. Eastwood, M.W. Farmen, P.W. Iversen, T.J. Craft, J.K. Smallwood, K.E. Garbison, N.W. Delapp, G.F. Smith. The minimum significant ratio: A statistical parameter to characterize the reproducibility of potency estimates from concentration-response assays and estimation by replicate-experiment studies. *Journal of Biomolecular Screening* 11 (2006) 253–61.

[40] C.J. Echeverri, P.A. Beachy, B. Baum, M. Boutros, F. Buchholz, S.K. Chanda, J. Downward, J. Ellenberg, A.G. Fraser, N. Hacohen, W.C. Hahn, A.L. Jackson, A. Kiger, P.S. Linsley, L. Lum, Y. Ma, B. Mathey-Prevot, D.E. Root, D.M. Sabatini, J. Taipale, N. Perrimon, R. Bernards. Minimizing the risk of reporting false positives in large-scale RNAi screens. *Nature Methods* 3 (2006) 777–9.

[41] C.J. Echeverri, N. Perrimon. High-throughput RNAi screening in cultured cells: a user's guide. *Nature Reviews Genetics* 7 (2006) 373–84.

[42] B. Efron. Large-scale simultaneous hypothesis testing: The choice of a null hypothesis. *Journal of the American Statistical Association* 99 (2004) 96–104.

[43] M. Eisenstein. Quality control. *Nature* 442 (2006) 1067–70.

[44] S.M. Elbashir, J. Harborth, W. Lendeckel, A. Yalcin, K. Weber, T. Tuschl. Duplexes of 21-nucleotide RNAs mediate RNA interference in cultured mammalian cells. *Nature* 411 (2001) 494–8.

[45] A.S. Espeseth, Q. Huang, A. Gates, M. Xu, Y. Yu, A.J. Simon, X.P. Shi, X.H.D. Zhang, P. Hodor, D.J. Stone, J. Burchard, G. Cavet, S. Bartz, P. Linsley, W.J. Ray, D. Hazuda. A genome wide analysis of ubiquitin ligases in APP processing identifies a novel regulator of BACE1 mRNA levels. *Molecular and Cellular Neuroscience* 33 (2006) 227–35.

[46] Y. Fedorov, E.M. Anderson, A. Birmingham, A. Reynolds, J. Karpilow, J. Robinson, D. Leake, W.S. Marshall, A. Khvorova. Off-target effects by siRNA can induce toxic phenotype. *Rna-A Publication of the Rna Society* 12 (2006) 1188–96.

[47] Y. Fedorov, A. King, E. Anderson, J. Karpilow, D. Ilsley, W. Marshall, A. Khvorova. Different delivery methods – Different expression profiles. *Nature Methods* 2 (2005) 241.

[48] A. Fire, S.Q. Xu, M.K. Montgomery, S.A. Kostas, S.E. Driver, C.C. Mello. Potent and specific genetic interference by double-stranded RNA in Caenorhabditis elegans. *Nature* 391 (1998) 806–11.

[49] R.A. Fisher. *Statistical Methods for Research Workers.* Edinburgh, Scotland, Oliver & Boyd, 1941.

[50] R.A. Fisher. *Statistical Methods for Research Workers.* Edinburgh, Scotland, Oliver & Boyd, 1925.

[51] E. Galun. *RNA Silencing.* Singapore, World Scientific, 2005.

[52] C. Genovese, L. Wasserman. Operating characteristics and extensions of the false discovery rate procedure. *Journal of the Royal Statistical Society Series B-Statistical Methodology* 64 (2002) 499–517.

[53] G.V. Glass. Primary, secondary, and meta-analysis of research. *Educational Researcher* 5 (1976) 3–8.

[54] T.R. Golub, D.K. Slonim, P. Tamayo, C. Huard, M. Gaasenbeek, J.P. Mesirov, H. Coller, M.L. Loh, J.R. Downing, M.A. Caligiuri, C.D. Bloomfield, E.S. Lander. Molecular classification of cancer: Class discovery and class prediction by gene expression monitoring. *Science* 286 (1999) 531–7.

[55] S. Goodman. A dirty dozen: Twelve P-value misconceptions. *Seminars in Hematology* 45 (2008) 135–40.

[56] P.J. Green, B.W. Silverman. *Nonparametric Regression and Generalized Linear Models: A Roughness Penalty Approach.* London, Chapman and Hall, 1994.

[57] Z. Guan, B.L. Wu, H.Y. Zhao. Nonparametric estimator of false discovery rate based on Bernstein polynomials. *Statistica Sinica* 18 (2008) 905–23.

[58] B. Gunter, C. Brideau, B. Pikounis, A. Liaw. Statistical and graphical methods for quality control determination of high-throughput screening data. *Journal of Biomolecular Screening* 8 (2003) 624–633.

[59] L. Guttman. What is not what in statistics. *Statistician* 26 (1977) 81–107.

[60] L. Guttman. The illogic of statistical inference for cumulative science. *Applied Stochastic Models and Data Analysis* 1 (1985) 3–10.

[61] I. Hamamoto, Y. Nishimura, T. Okamoto, H. Aizaki, M.Y. Liu, Y. Mori, T. Abe, T. Suzuki, M.M.C. Lai, T. Miyamura, K. Moriishi, Y. Matsuura. Human VAP-B is involved in hepatitis C virus replication through interaction with NS5A and NS5B. *Journal of Virology* 79 (2005) 13473–82.

[62] S.A. Haney. RNAi and high-content screening in target identification and validation. *Idrugs* 8 (2005) 997–1001.

[63] G.J. Hannon. RNA interference. *Nature* 418 (2002) 244–51.

[64] G.J. Hannon, P.D. Zamore. Small RNAs, big biology: biochemical studies of RNA interference. In: G.J. Hannon (Ed.), *A Guide to Gene Silencing.* New York, Cold Spring Harbor Laboratory Press, 2003, pp. 87–108.

[65] L.L. Harlow, S.A. Mulaik, J.H. Steiger. *What If There Were No Significance Tests?* Mahwah, NJ, Lawrence Erlbaum Associates, 1997.

[66] L.V. Hedges. Distribution theory for Glass's estimator of effect size and related estimators. *Journal of Educational Statistics* 6 (1981) 107–28.

[67] L.V. Hedges. I. Olkin Statistical Methods for Meta-Analysis. San Diego, CA, Academic Press, 1985.

[68]   M. Hollander, D. Wolfe. *Nonparametric Statistical Methods.* New York, John Wiley & Sons, 1999.

[69]   V. Hornung, M. Guenthner-Biller, C. Bourquin, A. Ablasser, M. Schlee, S. Uematsu, A. Noronha, M. Manoharan, S. Akira, A. de Fougerolles, S. Endres, G. Hartmann. Sequence-specific potent induction of IFN-alpha by short interfering RNA in plasmacytoid dendritic cells through TLR7. *Nature Medicine* 11 (2005) 263–70.

[70]   D. Howe, M. Costanzo, P. Fey, T. Gojobori, L. Hannick, W. Hide, D.P. Hill, R. Kania, M. Schaeffer, S. St Pierre, S. Twigger, O. White, S.Y. Rhee. Big data: The future of biocuration. *Nature* 455 (2008) 47–50.

[71]   P.J. Huber. *Robust Statistics.* New York, Wiley, 1981.

[72]   C.J. Huberty. A history of effect size indices. *Educational and Psychological Measurement* 62 (2002) 227–40.

[73]   C.J. Huberty, L.L. Lowman. Group overlap as a basis for effect size. *Educational and Psychological Measurement* 60 (2000) 543–63.

[74]   G. Hutvagner, J. McLachlan, A.E. Pasquinelli, E. Balint, T. Tuschl, P.D. Zamore. A cellular function for the RNA-interference enzyme Dicer in the maturation of the let-7 small temporal RNA. *Science* 293 (2001) 834–8.

[75]   J.P.A. Ioannidis. Microarrays and molecular research: noise discovery? *Lancet* 365 (2005) 454–5.

[76]   J.P.A. Ioannidis. Why most published research findings are false. *Plos Medicine* 2 (2005) 696–701.

[77]   P.W. Iversen, B.J. Eastwood, G.S. Sittampalam, K.L. Cox. A comparison of assay performance measures in screening assays: Signal window, Z factor, and assay variability ratio. *Journal of Biomolecular Screening* 11 (2006) 247–52.

[78]   A.L. Jackson, S.R. Bartz, J. Schelter, S.V. Kobayashi, J. Burchard, M. Mao, B. Li, G. Cavet, P.S. Linsley. Expression profiling reveals off-target gene regulation by RNAi. *Nature Biotechnology* 21 (2003) 635–7.

[79]   A.L. Jackson, J. Burchard, D. Leake, A. Reynolds, J. Schelter, J. Guo, J.M. Johnson, L. Lim, J. Karpilow, K. Nichols, W. Marshall, A. Khvorova, P.S. Linsley. Position-specific chemical modification of siRNAs reduces "off-target" transcript silencing. *RNA-A Publication of the RNA Society* 12 (2006) 1197–1205.

[80]   A.L. Jackson, P.S. Linsley. Noise amidst the silence: off-target effects of siRNAs? *Trends in Genetics* 20 (2004) 521–4.

[81]   A.D. Judge, V. Sood, J.R. Shaw, D. Fang, K. McClintock, I. MacLachlan. Sequence-dependent stimulation of the mammalian innate immune response by synthetic siRNA. *Nature Biotechnology* 23 (2005) 457–62.

[82]   P.D. Kassner. Discovery of novel targets with high throughput RNA interference screening. *Combinatorial Chemistry & High Throughput Screening* 11 (2008) 175–84.

[83]   D. Kevorkov, V. Makarenkov. Statistical analysis of systematic errors in high-throughput screening. *Journal of Biomolecular Screening* 10 (2005) 557–67.

[84]   R.E. Kirk. Practical significance: A concept whose time has come. *Educational and Psychological Measurement* 56 (1996) 746–59.

[85]   M.E. Kleinman, K. Yamada, A. Takeda, V. Chandrasekaran, M. Nozaki, J.Z. Baffi, R.J.C. Albuquerque, S. Yamasaki, M. Itaya, Y.Z. Pan, B. Appukuttan, D. Gibbs, Z.L. Yang, K. Kariko, B.K. Ambati, T.A. Wilgus, L.A. DiPietro, E. Sakurai, K. Zhang, J.R. Smith, E.W.

Taylor, J. Ambati. Sequence- and target-independent angiogenesis suppression by siRNA via TLR3. *Nature* 452 (2008) 591–7.

[86]   R.A. Klinghoffer, J. Frazier, J. Annis, J.D. Berndt, B.S. Roberts, W.T. Arthur, R. Lacson, X.H.D. Zhang, M. Ferrer, R.T. Moon, M.A. Cleary. A lentivirus-mediated genetic screen identifies dihydrofolate reductase (DHFR) as a modulator of b-actinin/GSK3 signaling. *PLoS ONE* 4 (2009) e6892.

[87]   R. Konig, C.Y. Chiang, B.P. Tu, S.F. Yan, P.D. DeJesus, A. Romero, T. Bergauer, A. Orth, U. Krueger, Y. Zhou, S.K. Chanda. A probability-based approach for the analysis of large-scale RNAi screens. *Nature Methods* 4 (2007) 847–9.

[88]   R. Konig, Y.Y. Zhou, D. Elleder, T.L. Diamond, G.M.C. Bonamy, J.T. Irelan, C.Y. Chiang, B.P. Tu, P.D. De Jesus, C.E. Lilley, S. Seidel, A.M. Opaluch, J.S. Caldwell, M.D. Weitzman, K.L. Kuhen, S. Bandyopadhyay, T. Ideker, A.P. Orth, L.J. Miraglia, F.D. Bushman, J.A. Young, S.K. Chanda. Global analysis of host-pathogen interactions that regulate early-stage HIV-1 replication. *Cell* 135 (2008) 49–60.

[89]   M. Lagos-Quintana, R. Rauhut, W. Lendeckel, T. Tuschl. Identification of novel genes coding for small expressed RNAs. *Science* 294 (2001) 853–8.

[90]   M. Langaas, B.H. Lindqvist, E. Ferkingstad. Estimating the proportion of true null hypotheses, with application to DNA microarray data. *Journal of the Royal Statistical Society Series B-Statistical Methodology* 67 (2005) 555–72.

[91]   N.C. Lau, L.P. Lim, E.G. Weinstein, D.P. Bartel. An abundant class of tiny RNAs with probable regulatory roles in Caenorhabditis elegans. *Science* 294 (2001) 858–62.

[92]   R.C. Lee, V. Ambros. An extensive class of small RNAs in Caenorhabditis elegans. *Science* 294 (2001) 862–4.

[93]   J.G. Liao, Y. Lin, Z.E. Selvanayagam, W.C.J. Shih. A mixture model for estimating the local false discovery rate in DNA microarray analysis. *Bioinformatics* 20 (2004) 2694–2701.

[94]   X.Y. Lin, X. Ruan, M.G. Anderson, J.A. McDowell, P.E. Kroeger, S.W. Fesik, Y. Shen. siRNA-mediated off-target gene silencing triggered by a 7 nt complementation. *Nucleic Acids Research* 33 (2005) 4527–35.

[95]   M. Llano, D.T. Saenz, A. Meehan, P. Wongthida, M. Peretz, W.H. Walker, W.L. Teo, E.M. Poeschla. An essential role for LEDGF/p75 in HIV integration. *Science* 314 (2006) 461–4.

[96]   I. Maeda, Y. Kohara, M. Yamamoto, A. Sugimoto. Large-scale analysis of gene function in Caenorhabditis elegans by high-throughput RNAi. *Current Biology* 11 (2001) 171–6.

[97]   G. Maertens, P. Cherepanov, W. Pluymers, K. Busschots, E. De Clercq, Z. Debyser, Y. Engelborghs. LEDGF/p75 is essential for nuclear and chromosomal targeting of HIV-1 integrase in human cells. *Journal of Biological Chemistry.* 278 (2003) 33528–39.

[98]   N. Mahanthappa. Translating RNA interference into therapies for human disease. *Pharmacogenomics* 6 (2005) 879–83.

[99]   N. Malo, J.A. Hanley, S. Cerquozzi, J. Pelletier, R. Nadon. Statistical practice in high-throughput screening data analysis. *Nature Biotechnology* 24 (2006) 167–75.

[100] J.C. Marioni, C.E. Mason, S.M. Mane, M. Stephens, Y. Gilad. RNA-seq: An assessment of technical reproducibility and comparison with gene expression arrays. *Genome Research* 18 (2008) 1509–17.

[101] R.S. Matson. *Applying Genomic and Proteomic Microarray Technology in Drug Discovery.* Boca Raton, FL,CRC Press, 2005.

[102] K.O. Mcgraw, S.P. Wong, A Common Language Effect Size Statistic. *Psychological Bulletin* 111 (1992) 361–365.

[103] G.J. McLachlan, R.W. Bean, L.B.T. Jones. A simple implementation of a normal mixture approach to differential gene expression in multiclass microarrays. *Bioinformatics* 22 (2006) 1608–15.

[104] J.E. McLean. *Improving Education through Action Research: A Guide for Administrators and Teachers.* Thousand Oaks, CA, Corwin Press, 1995.

[105] P.E. Meehl. Theory-testing in psychology and physics: A methodological paradox. *Philosophy of Science* 20 (1967) 103–15.

[106] P.E. Meehl. Theoretical risks and tabular asterisks – Karl, Ronald, and slow progress of soft psychology. *Journal of Consulting and Clinical Psychology* 46 (1978) 806–34.

[107] S. Michiels, S. Koscielny, C. Hill. Prediction of cancer outcome with microarrays: a multiple random validation strategy. *Lancet* 365 (2005) 488–92.

[108] P. Muller, D. Kuttenkeuler, V. Gesellchen, M.P. Zeidler, M. Boutros. Identification of JAK/STAT signalling components by genome-wide RNA interference. *Nature* 436 (2005) 871–5.

[109] C. Napoli, C. Lemieux, R. Jorgensen. Introduction of a chimeric chalcone synthase gene into petunia results in reversible co-suppression of homologous genes in trans. *Plant Cell* 2 (1990) 279–89.

[110] Neter J., M.H. Kutner, C.J. Nachtsheim. *Applied Linear Statistical Models.* Chicago, McGraw-Hill, 1996.

[111] J. Neyman, E.S. Pearson. On the use and interpretation of certain test criteria for purposes of statistical inference, Part I. *Biometrika* 29A (1928) 175–240.

[112] J. Neyman, E.S. Pearson. On the use and interpretation of certain test criteria for purposes of statistical inference, Part II. *Biometrika* 29A (1928) 263–94.

[113] K. Nybakken, S.A. Vokes, T.Y. Lin, A.P. McMahon, N. Perrimon. A genome-wide RNA interference screen in Drosophila melanogaster cells for new components of the Hh signaling pathway. *Nature Genetics* 37 (2005) 1323–32.

[114] A.J. Onwuegbuzie, J.R. Levin. Without supporting statistical evidence, where would reported measures of substantive importance lead? To no good effect. *Journal of Modern Applied Statistical Methods* 2 (2003) 133–51.

[115] D.B. Owen, K.J. Graswell, D.L. Hanson. Nonparametric upper confidence bounds for $P(Y < X)$ and confidence limits for $P(Y < X)$ when X and Y are normal. *Journal of the American Statistical Association* 59 (1964) 906–24.

[116] H.J. Park, E. Partridge, P. Cheung, J. Pawling, R. Donovan, J.L. Wrana, J.W. Dennis. Chemical enhancers of cytokine signaling that suppress microfilament turnover and tumor cell growth. *Cancer Research* 66 (2006) 3558–66.

[117] K. Pearson. On the correlation of characters not quantitatively measurable. *Philosophical Transactions of the Royal Society of London* 195 (1901) 1–47.

[118] L. Pelkmans, E. Fava, H. Grabner, M. Hannus, B. Habermann, E. Krausz, M. Zerial. Genome-wide analysis of human kinases in clathrin- and caveolae/raft-mediated endocytosis. *Nature* 436 (2005) 78–86.

[119] P. Pollard. How significant is "significance"? In: G. Keren and C. Lewis (Eds.), *A Handbook for Data Analysis in the Behavioral Sciences: Methodological Issues.* Hillsdale, NJ, Lawrence Erlbaum Associates, 1993, pp. 449–460.

[120] S. Pounds, S.W. Morris. Estimating the occurrence of false positives and false negatives in microarray studies by approximating and partitioning the empirical distribution of p-values. *Bioinformatics* 19 (2003) 1236–42.

[121] G. Randall, A. Grakoui, C.M. Rice. Clearance of replicating hepatitis C virus replicon RNAs in cell culture by small interfering RNAs. *Proceedings of the National Academy of Sciences of the United States of America* 100 (2003) 235–240.

[122] B. Reiser, I. Guttman. Statistical-Inference for Pr(Y-Less-Than-X) – the Normal Case. *Technometrics* 28 (1986) 253–7.

[123] D. Riester, F. Wirsching, G. Salinas, M. Keller, M. Gebinoga, S. Kamphausen, C. Merkwirth, R. Goetz, M. Wiesenfeldt, J. Sturzebecher, W. Bode, R. Friedrich, M. Thurk, A. Schwienhorst. Thrombin inhibitors identified by computer-assisted multiparameter design. Proceedings of the *National Academy of Sciences of the United States of America* 102 (2005) 8597–8602.

[124] S. Robin, F. Bar-Hen, J.J. Daudin, L. Pierre. A semi-parametric approach for mixture models: Application to local false discovery rate estimation. *Computational Statistics & Data Analysis* 51 (2007) 5483–93.

[125] D.M. Rom, E. Hwang. Testing for individual and population equivalence based on the proportion of similar responses. *Statistics in Medicine* 15 (1996) 1489–1505.

[126] R. Rosenthal, R.L. Rosnow, D.B. Rubin. *Contrasts and Effect Sizes in Behavioral Research.* Cambridge, UK, Cambridge University Press, 2000.

[127] W.W. Rozeboom. The fallacy of the null-hypothesis significance test. Psychological Bulletin 57 (1960) 416–28.

[128] G. Ruvkun. Molecular biology – Glimpses of a tiny RNA world. *Science* 294 (2001) 797–9.

[129] F.E. Satterthwaite. An approximate distribution of estimates of variance components. *Biometrics Bulletin* 2 (1946) 110–14.

[130] P.C. Scacheri, O. Rozenblatt-Rosen, N.J. Caplen, T.G. Wolfsberg, L. Umayam, J.C. Lee, C.M. Hughes, K.S. Shanmugam, A. Bhattacharjee, M. Meyerson, F.S. Collins. Short interfering RNAs can induce unexpected and divergent changes in the levels of untargeted proteins in mammalian cells. *Proceedings of the National Academy of Sciences of the United States of America* 101 (2004) 1892–7.

[131] F.L. Schmidt. What do data really mean – Research findings, metaanalysis, and cumulative knowledge in psychology. *American Psychologist* 47 (1992) 1173–81.

[132] F.L. Schmidt. Statistical significance testing and cumulative knowledge in psychology: Implications for training of researchers. *Psychological Methods* 1 (1996) 115–29.

[133] D. Semizarov, L. Frost, A. Sarthy, P. Kroeger, D.N. Halbert, S.W. Fesik. Specificity of short interfering RNA determined through gene expression signatures. *Proceedings of the National Academy of Sciences of the United States of America* 100 (2003) 6347–52.

[134] S. Senn, Testing for individual and population equivalence based on the proportion of similar responses. *Statistics in Medicine* 16 (1997) 1303–1305.

[135] S.S. Shapiro, M.B. Wilk. An analysis of variance test for normality (complete samples). *Biometrika* 52 (1965) 591–611.

[136] J. Shoemaker. Statistical challenges with gene expression studies. *Pharmacogenomics* 7 (2006) 511–19.

[137] O. Snove, T. Holen. Many commonly used siRNAs risk off-target activity. *Biochemical and Biophysical Research Communications* 319 (2004) 256–63.

[138] B. Sonnichsen, L.B. Koski, A. Walsh, P. Marschall, B. Neumann, M. Brehm, A.M. Alleaume, J. Artelt, P. Bettencourt, E. Cassin, M. Hewitson, C. Holz, M. Khan, S. Lazik, C. Martin, B. Nitzsche, M. Ruer, J. Stamford, M. Winzi, R. Heinkel, M. Roder, J. Finell, H. Hantsch, S.J.M. Jones, M. Jones, F. Piano, K.C. Gunsalus, K. Oegema, P. Gonczy, A. Coulson, A.A. Hyman, C.J. Echeverri. Full-genome RNAi profiling of early embryogenesis in Caenorhabditis elegans. *Nature* 434 (2005) 462–9.

[139] J.H. Steiger. Beyond the F test: Effect size confidence intervals and tests of close fit in the analysis of variance and contrast analysis. *Psychological Methods* 9 (2004) 164–82.

[140] M.A. Stephens. EDF statistics for goodness of fit and some comparisons. *Journal of the American Statistical Association* 69 (1974) 730–7.

[141] J.A.C. Sterne, G.D. Smith. Sifting the evidence – what's wrong with significance tests? *British Medical Journal* 322 (2001) 226–31.

[142] R.A. Stine, J.F. Heyse. Non-parametric estimates of overlap. *Statistics in Medicine* 20 (2001) 215–36.

[143] D.J. Stone, S. Marine, J. Majercak, W.J. Ray, A. Espeseth, A. Simon, M. Ferrer. High-throughput screening by RNA interference – Control of two distinct types of variance. *Cell Cycle* 6 (2007) 898–901.

[144] J.D. Storey. A direct approach to false discovery rates. *Journal of the Royal Statistical Society Series B-Statistical Methodology* 64 (2002) 479–98.

[145] J.D. Storey. The positive false discovery rate: A Bayesian interpretation and the q-value. *The Annals of Statistics* 31 (2003) 2013–35.

[146] J.D. Storey, R. Tibshirani. Statistical significance for genomewide studies. *Proceedings of the National Academy of Sciences of the United States of America* 100 (2003) 9440–5.

[147] K. Strimmer. A unified approach to false discovery rate estimation. *BMC Bioinformatics* 9 (2008) 303.

[148] Y.X. Sui, Z.J. Wu. Alternative statistical parameter for high-throughput screening assay quality assessment. *Journal of Biomolecular Screening* 12 (2007) 229–34.

[149] A.C. Tamhane, D.D. Dunlop. *Statistics and Data Analysis.* Upper Saddle River, NJ, Prentice-Hall, 2000.

[150] H. Tian, L.C. Cao, Y.P. Tan, S. Williams, L.L. Chen, T. Matray, A. Chenna, S. Moore, V. Hernandez, V. Xiao, M.X. Tang, S. Singh. Multiplex mRNA assay using electrophoretic tags for high-throughput gene expression analysis. *Nucleic Acids Research* 32 (2004) e126.

[151] N.H. Timm. Estimating effect sizes in exploratory experimental studies when using a linear model. *American Statistician* 58 (2004) 213–17.

[152] C.A. Tsai, H.M. Hsueh, J.J. Chen. Estimation of false discovery rates in multiple testing: application to gene microarray data. *Biometrics* 59 (2003) 1071–81.

[153] T. Vacha-Haase, B. Thompson. How to estimate and interpret various effect sizes. *Journal of Counseling Psychology* 51 (2004) 473–81.

[154] W.N. Venables, B.D. Ripley. *Modern Applied Statistics with S.* New York, Springer, 2002.

[155] D.F. Vysochanskij, Y.I. Petunin. Justification of the 3-sigma rule for unimodal distribution. *Theory of Probability and Mathematical Statistics* 21 (1980) 25–36.

[156] J.D. Watson, F.H. Crick. Molecular structure of nucleic acids; a structure for deoxyribose nucleic acid. *Nature* 171 (1953) 737–8.

[157] F. Wilcoxon. Individual comparisons by ranking methods. *Biometrics* 1 (1945) 80–3.

[158] A.T. Willingham, Q.L. Deveraux, G.M. Hampton, P. Aza-Blanc. RNAi and HTS: exploring cancer by systematic loss-of-function. *Oncogene* 23 (2004) 8392–8400.

[159] J.H. Zhang, T.D.Y. Chung, K.R. Oldenburg. A simple statistical parameter for use in evaluation and validation of high throughput screening assays. *Journal of Biomolecular Screening* 4 (1999) 67–73.

[160] J.H. Zhang, T.D.Y. Chung, K.R. Oldenburg. Confirmation of primary active substances from high throughput screening of chemical and biological populations: A statistical approach and practical considerations. *Journal of Combinatorial Chemistry* 2 (2000) 258–65.

[161] X.H.D. Zhang. A new method with flexible and balanced control of false negatives and false positives for hit selection in RNA interference high-throughput screening assays. *Journal of Biomolecular Screening* 12 (2007) 645–55.

[162] X.H.D. Zhang. A pair of new statistical parameters for quality control in RNA interference high-throughput screening assays. *Genomics* 89 (2007) 552–61.

[163] X.H.D. Zhang. *New concept of contrast in statistical analysis.* The American Statistical Association Proceedings, Alexandria, VA, 2007, pp. 637–648.

[164] X.H.D. Zhang. Threshold determination of strictly standardized mean difference in RNA interference high throughput screening assays. *IMECS Proceeding* (2007) 261–6.

[165] X.H.D. Zhang. Genome-wide screens for effective siRNAs through assessing the size of siRNA effects. *BMC Research Notes* 1 (2008) 33.

[166] X.H.D. Zhang. Novel analytic criteria and effective plate designs for quality control in genome-scale RNAi screens. *Journal of Biomolecular Screening* 13 (2008) 363–77.

[167] X.H.D. Zhang. A method for effectively comparing gene effects in multiple conditions in RNAi and expression-profiling research. *Pharmacogenomics* 10 (2009) 345–58.

[168] X.H.D. Zhang. *A statistical method assessing collective activity of multiple siRNAs targeting a gene in RNAi screens.* Alexandria, VA. The American Statistical Association Proceedings. 2010 (In press).

[169] X.H.D. Zhang. An effective method for controlling false discovery and false non-discovery rates in genome-scale RNAi screens. *Journal of Biomolecular Screening* 15 (2010) 1116–22.

[170] X.H.D. Zhang. Assessing the size of gene or RNAi effects in multifactor high-throughput experiments. *Pharmacogenomics* 11 (2010) 199–213.

[171] X.H.D. Zhang. Strictly standardized mean difference, standardized mean difference and classical t-test for the comparison of two groups. *Statistics in Biopharmaceutical Research* 2 (2010) 292–9.

[172] X.H.D. Zhang, A. Espeseth, N. Chung, M. Ferrer. *Evaluation of a novel metric for quality control in an RNA interference high-throughput screening assay.* Las Vegas, NV, CSREA Press, 2006, pp. 385–390.

[173] X.H.D. Zhang, A.S. Espeseth, E.N. Johnson, J. Chin, A. Gates, L.J. Mitnaul, S.D. Marine, J. Tian, E.M. Stec, P. Kunapuli, D.J. Holder, J.F. Heyse, B. Strulovici, M. Ferrer. Integrating experimental and analytic approaches to improve data quality in genome-wide RNAi screens. *Journal of Biomolecular Screening* 13 (2008) 378–89.

[174] X.H.D. Zhang, M. Ferrer, A.S. Espeseth, S.D. Marine, E.M. Stec, M.A. Crackower, D.J. Holder, J.F. Heyse, B. Strulovici. The use of strictly standardized mean difference for hit selection in primary RNA interference high-throughput screening experiments. *Journal of Biomolecular Screening* 12 (2007) 497–509.

[175] X.H.D. Zhang, J.F. Heyse. Determination of sample size in genome-scale RNAi screens. *Bioinformatics* 25 (2009) 841–4.

[176] X.H.D. Zhang, P.F. Kuan, M. Ferrer, X.H. Shu, Y.C.X. Liu, A.T. Gates, P. Kunapuli, E.M. Stec, M. Xu, S.D. Marine, D.J. Holder, B. Strulovici, J.F. Heyse, A.S. Espeseth. Hit selection with false discovery rate control in genome-scale RNAi screens. *Nucleic Acids Research* 36 (2008) 4667–79.

[177] X.H.D. Zhang, R. Lacson, R. Yang, S. Marine, A. McCampbell, D. Toolan, T. Hare, J. Kajdas, D.J. Holder, J.F. Heyse, M. Ferrer. The use of SSMD-based false discovery and false non-discovery rates in genome-scale RNAi screens. *Journal of Biomolecular Screening* 15 (2010) 1123–31.

[178] X.H.D. Zhang, S.D. Marine, M. Ferrer. Error rates and powers in genome-scale RNAi screens. *Journal of Biomolecular Screening* 14 (2009) 230–8.

[179] X.H.D. Zhang, X.C. Yang, N. Chung, A. Gates, E. Stec, P. Kunapuli, D.J. Holder, M. Ferrer, A. Espeseth. Exploring statistical methods for hit selection in RNA interference high throughput screening experiments. *The American Statistical Association Proceedings*, 2005, pp. 775–80.

[180] X.H.D. Zhang, X.C. Yang, N.J. Chung, A. Gates, E. Stec, P. Kunapuli, D.J. Holder, M. Ferrer, A.S. Espeseth. Robust statistical methods for hit selection in RNA interference high-throughput screening experiments. *Pharmacogenomics* 7 (2006) 299–309.

[181] H.F. Zhao, D. L'Abbe, N. Jolicoeur, M.Q. Wu, Z. Li, Z.B. Yu, S.H. Shen. High-throughput screening of effective siRNAs from RNAi libraries delivered via bacterial invasion. *Nature Methods* 2 (2005) 967–73.

[182] W.Q. Zhao, F. Santini, R. Breese, D. Ross, X.H.D. Zhang, D.J. Stone, M. Ferrer, M. Townsend, A.L. Wolfe, M.A. Seager, G.G. Kinney, P.J. Shughrue, W.J. Ray. Inhibition of calcineurin-mediated endocytosis and alpha-amino-3-hydroxy-5-methyl-4-isoxazolepropionic acid (AMPA) receptors prevents amyloid beta oligomer-induced synaptic disruption. *Journal of Biological Chemistry* 285 (2010) 7619–32.

[183] H.L. Zhou, M. Xu, Q. Huang, A.T. Gates, X.H.D. Zhang, J.C. Castle, E. Stec, M. Ferrer, B. Strulovici, D.J. Hazuda, A.S. Espeseth. Genome-scale RNAi screen for host factors required for HIV replication. *Cell Host & Microbe* 4 (2008) 495–504.

[184] W. Zhou. Statistical inference for $P(X < Y)$. *Statistics in Medicine* 27 (2008) 257–79.

[185] P. Zuck, E.M. Murray, E. Stec, J.A. Grobler, A.J. Simon, B. Strulovici, J. Inglese, O.A. Flores, M. Ferrer. A cell-based beta-lactamase reporter gene assay for the identification of inhibitors of hepatitis C virus replication. *Analytical Biochemistry* 334 (2004) 344–55.

[186] [Anon]. A billion dollar punt. *Nature Biotechnology* 24 (2006) 1453.

Index